BTEC National
Construction

Study Guide

A PEARSON COMPANY

Published by
Pearson Education Ltd
Edinburgh Gate
Harlow
Essex CM20 2JE

© Pearson Education Ltd 2008

First published 2008

ISBN 978-1-84690-217-8

Typeset by Techset Composition Ltd
Printed in Great Britain by Henry Ling Ltd, at the Dorset Press, Dorchester, Dorset

Cover image © Volker Kreinacke/iStock

The publisher's policy is to use paper manufactured from sustainable forests.

All reasonable efforts have been made to trace and contact original copyright holders.

This material offers high-quality support for the delivery of Edexcel qualifications. This does not mean that it is essential to use it to achieve any Edexcel qualification, nor does it mean that this is the only suitable material available to support any Edexcel qualification. No Edexcel-published material will be used verbatim in setting any Edexcel assessment and any resource lists produced by Edexcel shall include this and other appropriate texts.

Acknowledgements
We are grateful to the following for permission to reproduce copyrighted material:
p.104 Health and Safety Executive for an extract from "Construction Deaths Down in 2004/2005 – But not a time to be complacent" published on www.hse.gov.uk 29 July 2005 © Crown copyright 2008; p.135 The National Learning Network for details from www.nln.ac.uk copyright © The Learning and Skills Council; Microsoft for Microsoft product screenshots, reprinted with permission from Microsoft Corporation.

Photographs
p.142 © Ken Price/Construction Photography; p.143 © plus49/Construction Photography.

CONTENTS

PREFACE

If you've already followed a BTEC First programme, you will know that this is an exciting way to study; if you're fresh from GCSEs you will find that from now on you will be in charge of your own learning. This guide has been written specially for you, to help you get started and then succeed on your BTEC National course.

The **Introduction** concentrates on making sure you have all the right facts about your course at your fingertips. Also, it guides you through the important skills you need to develop if you want to do well, including:

- managing your time
- researching information
- preparing a presentation.

Keep this by your side throughout your course and dip into it whenever you need to.

The **Activities** give you tasks to do on your own, in a small group or as a class. They will help you internalise your learning and then prepare you for assessment by practising your skills and showing how much you know. These activities are not for assessment.

The sample **Marked Assignment** shows you what other students have done to gain a Pass, Merit or Distinction. By seeing what past students have done, you should be able to improve your own grade.

Your BTEC National will cover 6, 12 or 18 units depending on whether you are doing an Award, Certificate or Diploma. In this guide the activities cover sections from Unit 2 – Construction and the Environment, Unit 3 – Mathematics in Construction and the Built Environment, and Unit 7 – Planning, Organisation and Control of Resources in Construction and the Built Environment. The marked assignment covers work produced for Unit 1 – Health, Safety and Welfare in Construction and the Built Environment.

Because the guide covers only four units, it is essential that you do all the other work your tutors set you. You will have to research information in textbooks, in the library and on the internet. You should have the opportunity to visit local organisations and welcome visiting speakers to your institution. This is a great way to find out more about your chosen vocational area – the type of jobs that are available and what the work is really like.

This Guide is a taster, an introduction to your BTEC National. Use it as such and make the most of the rich learning environment that your tutors will provide for you. Your BTEC National will give you an excellent base for further study, a broad understanding of construction, and the knowledge you need to succeed in the world of work. Remember, thousands of students have achieved a BTEC National and are now studying for a degree or are at work, building a successful career.

INTRODUCTION

SEVEN STEPS TO SUCCESS ON YOUR BTEC NATIONAL

You have received this guide because you have decided to do a BTEC National qualification. You may even have started your course. At this stage you should feel good about your decision. BTEC Nationals have many benefits – they are well-known and respected qualifications, they provide excellent preparation for future work or help you to get into university if that is your aim. If you are already at work then gaining a BTEC National will increase your value to your employer and help to prepare you for promotion.

Despite all these benefits though, you may be rather apprehensive about your ability to cope. Or you may be wildly enthusiastic about the whole course! More probably, you are somewhere between the two – perhaps quietly confident most of the time but sometimes worried that you may get out of your depth as the course progresses. You may be certain you made the right choice or still have days when your decision worries you. You may understand exactly what the course entails and what you have to do – or still feel rather bewildered, given all the new stuff you have to get your head around.

Your tutors will use the induction sessions at the start of your course to explain the important information they want you to know. At the time, though, it can be difficult to remember everything. This is especially true if you have just left school and are now studying in a new environment, among a group of people you have only just met. It is often only later that you think of useful questions to ask. Sometimes, misunderstandings or difficulties may only surface weeks or months into a course – and may continue for some time unless they are quickly resolved.

Make sure you have all the right facts

This student guide has been written to help to minimise these difficulties, so that you get the most out of your BTEC National course from day one. You can read through it at your own pace. You can look back at it whenever you have a problem or query.

This Introduction concentrates on making sure you have all the right facts about your course at your fingertips. This includes a **Glossary** (on page 32) which explains the specialist terms you may hear or read – including words and phrases highlighted in bold type in this Introduction.

The Introduction also guides you through the important skills you need to develop if you want to do well – such as managing your time, researching information and preparing a presentation; as well as reminding you about the key skills you will need to do justice to your work, such as good written and verbal communications.

- Use the PlusPoint boxes in each section to help you to stay focused on the essentials.

- Use the Action Point boxes to check out things you need to know or do right now.

- Refer to the Glossary (on page 32) if you need to check the meaning of any of the specialist terms you may hear or read.

Remember, thousands of students have achieved BTEC National Diplomas and are now studying for a degree or at work, building a successful career. Many were nervous and unsure of themselves at the outset – and very few experienced absolutely no setbacks during the course. What they did have, though, was a belief in their own ability to do well if they concentrated on getting things right one step at a time. This Introduction enables you to do exactly the same!

STEP ONE

UNDERSTAND YOUR COURSE AND HOW IT WORKS

What is a BTEC qualification and what does it involve? What will you be expected to do on the course? What can you do afterwards? How does this National differ from 'A' levels or a BTEC First qualification?

All these are common questions – but not all prospective students ask them! Did you? And, if so, did you really listen to the answers? And can you remember them now?

If you have already completed a BTEC First course then you may know some of the answers – although you may not appreciate some of the differences between that course and your new one.

Let's start by checking out the basics.

- All BTEC National qualifications are **vocational** or **work-related**. This doesn't mean that they give you all the skills that you need to do a job. It does mean that you gain the specific knowledge and understanding relevant to your chosen subject or area of work. This means that when you start in a job you will learn how to do the work more quickly and should progress further. If you are already employed, it means you become more valuable to your employer. You can choose to study a BTEC National in a wide range of vocational areas, such as Business, Health and Social Care, IT, Performing Arts and many others.

- There are three types of BTEC National qualification and each has a different number of units.

  - The BTEC National Award usually has 6 units and takes 360 **guided learning hours (GLH)** to complete. It is often offered as a part-time or short course but you may be one of the many students doing an Award alongside A-levels as a full-time course. An Award is equivalent to one 'A' level.

  - The BTEC National Certificate usually has 12 units and takes 720 GLH to complete. You may be able to study for the Certificate on a part-time or full-time course. It is equivalent to two 'A' levels.

– The BTEC National Diploma usually has 18 units and takes 1080 GLH to complete. It is normally offered as a two-year full-time course. It is equivalent to three 'A' levels.

These qualifications are often described as **nested**. This means that they fit inside each other (rather like Russian dolls!) because the same units are common to them all. This means that if you want to progress from one to another you can do so easily by simply completing more units.

- Every BTEC National qualification has a set number of **core units**. These are the compulsory units every student must complete. The number of core units you will do on your course depends upon the vocational area you are studying.

- All BTEC National qualifications also have a range of **specialist units** from which you may be able to make a choice. These enable you to study particular areas in more depth.

- Some BTEC National qualifications have **specialist core units**. These are mandatory units you will have to complete if you want to follow a particular pathway in certain vocational areas. Engineering is an example of a qualification with the over-arching title, Engineering, which has a set of core units that all students must complete. Then, depending what type of engineering a student wants to follow, there are more specialist core units that must be studied.

- On all BTEC courses you are expected to be in charge of your own learning. If you have completed a BTEC First, you will already have been introduced to this idea, but you can expect the situation to be rather different now that you are working at BTEC National level. Students on a BTEC First course will be expected to need more guidance whilst they develop their skills and find their feet. In some cases, this might last quite some time. On a BTEC National course you will be expected to take more responsibility for yourself and your own learning almost from the outset. You will quickly be expected to start thinking for yourself. This means planning what to do and carrying out a task without needing constant reminders. This doesn't mean that your tutor won't give you help and guidance when you need it. It does mean, though, that you need to be 'self-starting' and to be able to use your own initiative. You also need to be able to assess your own performance and make improvements when necessary. If you enjoy having the freedom to make your own decisions and work at your own pace then you will welcome this type of learning with open arms. However, there are dangers! If you are a procrastinator (look up this word if you don't know what it means!) then it's quite likely you will quickly get in a muddle. In this case read Step 3 – Use your time wisely – very carefully indeed!

- The way you are assessed and graded on a BTEC course is different from an 'A' level course, although you will still obtain UCAS points which you need if you want to go to university. You can read about this in the next section.

PLUSPOINTS

+ You can usually choose to study part-time or full-time for your BTEC National and do an Award, Certificate or Diploma and progress easily from one to the other.

+ You will study both core units and specialist units on your course.

+ When you have completed your BTEC course you can get a job (or **apprenticeship**), use your qualification to develop your career and/or continue your studies to degree level.

+ You are responsible for your own learning on a BTEC course. This prepares you for life at work or at university when you will be expected to be self-starting and to use your own initiative.

ACTION POINTS

✓ Check you know whether you are studying for an Award, Certificate or Diploma and find out the number of units you will be studying for your BTEC National qualification.

✓ Find out which are core and which are specialist units, and which specialist units are offered at your school or college.

✓ Check out the length of your course and when you will be studying each unit.

✓ Explore the Edexcel website at www.edexcel.org.uk. Your first task is to find what's available for your particular BTEC National qualification. Start by finding National qualifications, then look for your vocational area and check you are looking at the 2007 schemes. Then find the specification for your course. Don't print this out – it is far too long. You could, of course, save it if you want to refer to it regularly or you could just look through it for interest and then bookmark the pages relating to your qualification for future reference.

✓ Score yourself out of 5 (where 0 is awful and 5 is excellent) on each of the following to see how much improvement is needed for you to become responsible for your own learning!

Being punctual; organisational ability; tidiness; working accurately; finding and correcting own mistakes; solving problems; accepting responsibility; working with details; planning how to do a job; using own initiative; thinking up new ideas; meeting deadlines.

✓ Draw up your own action plan to improve any areas where you are weak. Talk this through at your next individual **tutorial**.

STEP TWO

UNDERSTAND HOW YOU ARE ASSESSED AND GRADED – AND USE THIS KNOWLEDGE TO YOUR ADVANTAGE!

If you already have a BTEC First qualification, you may think that you don't need to read this section because you assume that BTEC National is simply more of the same. Whilst there are some broad similarities, you will now be working at an entirely different level and the grades you get for your work could be absolutely crucial to your future plans.

Equally, if you have opted for BTEC National rather than 'A' level because you thought you would have less work (or writing) to do then you need to read this section very carefully. Indeed, if you chose your BTEC National because you thought it would guarantee you an easy life, you are likely to get quite a shock when reality hits home!

It is true that, unlike 'A' levels, there are no exams on a BTEC course. However, to do well you need to understand the importance of your assignments, how these are graded and how these convert into unit points and UCAS points. This is the focus of this section.

Your assignments

On a BTEC National course your learning is assessed by means of **assignments** set by your tutors and given to you to complete throughout your course.

- Your tutors will use a variety of **assessment methods**, such as case studies, projects, presentations and shows to obtain evidence of your skills and knowledge to date. You may also be given work-based or **time-constrained** assignments – where your performance might be observed and assessed. It will depend very much on the vocational area you are studying.

- Important skills you will need to learn are how to research information (see page 25) and how to use your time effectively, particularly if you have to cope with several assignments at the same time (see page 12). You may also be expected to work cooperatively as a member of a team to complete some parts of your assignments – especially if you are doing a subject like Performing Arts – or to prepare a presentation (see page 26).

- All your assignments are based on **learning outcomes** set by Edexcel. These are listed for each unit in your course specification. You have to meet *all* the learning outcomes to pass the unit.

Your grades

On a BTEC National course, assignments that meet the learning outcomes are graded as Pass, Merit or Distinction.

- The difference between these grades has very little to do with how much you write! Edexcel sets out the **grading criteria** for the different grades in a **grading grid**. This identifies the **higher-level skills** you have to demonstrate to earn a higher grade. You can find out more about this, and read examples of good (and not so good) answers to assignments at Pass, Merit and Distinction level in the marked assignments section starting on page 157. You will also find out more about getting the best grade you can in Step 5 – Understand your assessment – on page 16.

- Your grades for all your assignments earn you **unit points**. The number of points you get for each unit is added together and your total score determines your final grade(s) for the qualification – again either Pass, Merit or Distinction. You get one final grade if you are taking a BTEC National Award, two if you are taking a BTEC National Certificate and three if you are taking a BTEC National Diploma.

- Your points and overall grade(s) also convert to **UCAS points** which you will need if you want to apply to study on a degree course. As an example, if you are studying a BTEC National Diploma, and achieve three final pass grades you will achieve 120 UCAS points. If you achieve three final distinction grades the number of UCAS points you have earned goes up to 360.

- It is important to note that you start earning both unit and UCAS points from the very first assignment you complete! This means that if you take a long time to settle into your course, or to start working productively, you could easily lose valuable points for quite some time. If you have your heart set on a particular university or degree course then this could limit your choices. Whichever way you look at it, it is silly to squander potentially good grades for an assignment and their equivalent points, just because you didn't really understand what you had to do – which is why this guide has been written to help you!

- If you take a little time to understand how **grade boundaries** work, you can see where you need to concentrate your efforts to get the best final grade possible. Let's give a simple example. Chris and Shaheeda both want to go to university and have worked hard on their BTEC National Diploma course. Chris ends with a total score of 226 unit points which converts to 280 UCAS points. Shaheeda ends with a total score of 228 unit points – just two points more – which converts to 320 UCAS points! This is because a score of between 204 and 227 unit points gives 280 UCAS points, whereas a score of 228 – 251 points gives 320 UCAS points. Shaheeda is pleased because this increases her chances of getting a place on the degree course she wants. Chris is annoyed. He says if he had known then he would have put more effort into his last assignment to get two points more.

■ It is always tempting to spend time on work you like doing, rather than work you don't – but this can be a mistake if you have already done the best you can at an assignment and it would already earn a very good grade. Instead you should now concentrate on improving an assignment which covers an area where you know you are weak, because this will boost your overall grade(s). You will learn more about this in Step 3 – Use your time wisely.

PLUSPOINTS

+ Your learning is assessed in a variety of ways, such as by assignments, projects and case studies. You will need to be able to research effectively, manage your own time and work well with other people to succeed.

+ You need to demonstrate specific knowledge and skills to achieve the learning outcomes set by Edexcel. You need to demonstrate you can meet all the learning outcomes to pass a unit.

+ Higher-level skills are required for higher grades. The grading criteria for Pass, Merit and Distinction grades are set out in a grading grid for the unit.

+ The assessment grades of Pass, Merit and Distinction convert to unit points. The total number of unit points you receive during the course determines your final overall grade(s) and the UCAS points you have earned.

+ Working effectively from the beginning maximises your chances of achieving a good qualification grade. Understanding grade boundaries enables you to get the best final grade(s) possible.

ACTION POINTS

✓ Find the learning outcomes for the units you are currently studying. Your tutor may have given you these already, or you can find them in the specification for your course that you already accessed at www.edexcel.org.uk.

✓ Look at the grading grid for the units and identify the way the evidence required changes to achieve the higher grades. Don't worry if there are some words that you do not understand – these are explained in more detail on page 32 of this guide.

✓ If you are still unsure how the unit points system works, ask your tutor to explain it to you.

✓ Check out the number of UCAS points you would need for any course or university in which you are interested.

✓ Keep a record of the unit points you earn throughout your course and check regularly how this is affecting your overall grade(s), based on the grade boundaries for your qualification. Your tutor will give you this information or you can check it yourself in the specification for your course on the Edexcel website.

STEP THREE

USE YOUR TIME WISELY

Most students on a BTEC National course are trying to combine their course commitments with a number of others – such as a job (either full or part-time) and family responsibilities. In addition, they still want time to meet with friends, enjoy a social life and keep up hobbies and interests that they have.

Starting the course doesn't mean that you have to hide away for months if you want to do well. It does mean that you have to use your time wisely if you want to do well, stay sane and keep a balance in your life.

You will only do this if you make time work for you, rather than against you, by taking control. This means that you decide what you are doing, when you are doing it and work purposefully; rather than simply reacting to problems or panicking madly because you've yet another deadline staring you in the face.

Use your time wisely

This becomes even more important as your course progresses because your workload is likely to increase, particularly towards the end of a term. In the early days you may be beautifully organised and able to cope easily. Then you may find you have several tasks to complete simultaneously as well as some research to start. Then you get two assignments in the same week from different tutors – as well as having a presentation to prepare. Then another assignment is scheduled for the following week – and so on. This is not because your tutors are being deliberately difficult. Indeed, most will try to schedule your assignments to avoid such clashes. The problem, of course, is that none of your tutors can assess your abilities until you have learned something – so if several units start and end at the same time it is highly likely there will be some overlap between your assignments.

To cope when the going gets tough, without collapsing into an exhausted heap, you need to learn a few time management skills.

- **Pinpoint where your time goes at the moment** Time is like money – it's usually difficult to work out where it all went! Work out how much time you currently spend at college, at work, at home and on social activities. Check, too, how much time you waste each week – and why this happens. Are you disorganised or do you easily get distracted? Then identify commitments that are vital and those that are optional so that you know where you can find time if you need to.

- **Plan when and where to work** It is unrealistic not to expect to do quite a lot of work for your course in your own time. It is also better to work regularly, and in relatively short bursts, than to work just once or twice a week for very long stretches. In addition to deciding when to work, and for how long, you also need to think about when and where to work. If you are a lark, you will work better early in the day; if you are an owl, you will be at your best later on. Whatever time you work, you need somewhere quiet so that you can concentrate and with space for books and other resources you need. If the words 'quiet oasis' and 'your house' are totally incompatible at any time of the day or night

11

then check out the opening hours of your local and college library so that you have an escape route if you need it. If you are trying to combine studying with parental responsibilities it is sensible to factor in your children's commitments – and work around their bedtimes too! Store up favours, too, from friends and grandparents that you can call in if you get desperate for extra time when an assignment deadline is looming.

- **Schedule your commitments** Keep a diary or (even better) a wall chart and write down every appointment you make or task you are given. It is useful to use a colour code to differentiate between personal and work or course commitments. You may also want to enter assignment review dates with your tutor in one colour and final deadline dates in another. Keep your diary or chart up-to-date by adding any new dates promptly every time you receive another task or assignment or whenever you make any other arrangements. Keep checking ahead so that you always have prior warning when important dates are looming. This stops you from planning a heavy social week when you will be at your busiest at work or college and from arranging a dental appointment on the morning when you and your team are scheduled to give an important presentation!

- **Prioritise your work** This means doing the most important and urgent task first, rather than the one you like the most! Normally this will be the task or assignment with the nearest deadline. There are two exceptions. Sometimes you may need to send off for information and allow time for it to arrive. It is therefore sensible to do this first so that you are not held up later. The second is when you have to take account of other people's schedules – because you are working in a team or are arranging to interview someone, for example. In this case you will have to arrange your schedule around their needs, not just your own.

- **Set sensible timescales** Trying to do work at the last minute or in a rush is never satisfactory, so it is wise always to allocate more time than you think you will need, never less. Remember, too, to include all the stages of a complex task or assignment, such as researching the information, deciding what to use, creating a first draft, checking it and making improvements and printing it out. If you are planning to do any of your work in a central facility always allow extra time and try to start work early. If you arrive at the last minute you may find every computer and printer is fully utilised until closing time.

- **Learn self-discipline!** This means not putting things off (procrastinating!) because you don't know where to start or don't feel in the mood. Unless you are ill, you have to find some way of persuading yourself to work. One way is to bribe yourself. Make a start and promise yourself that if you work productively for 30 minutes then you deserve a small reward. After 30 minutes you may have become more engrossed and want to keep going a little longer. Otherwise at least you have made a start, so it's easier to come back and do more later. It doesn't matter whether you have research to do, an assignment to write up, a coaching session to plan, or lines to learn, you need to be self-disciplined.

- **Take regular breaks and keep your life in balance** Don't go to the opposite extreme and work for hours on end. Take regular breaks to give yourself a rest and a change of activity. You need to recharge your batteries! Similarly, don't cancel every social arrangement so that you can work 24/7. Whilst this may be occasionally necessary if you have several deadlines looming simultaneously, it should only be a last resort. If you find yourself doing this regularly then go back to the beginning of this section and see where your time-management planning is going wrong.

PLUSPOINTS

+ Being in control of your time enables you to balance your commitments according to their importance and allows you not let to anyone down – including yourself.

+ Controlling time involves knowing how you spend (and waste!) your time now, planning when best to do work, scheduling your commitments and setting sensible timescales for work to be done.

+ Knowing how to prioritise means that you will schedule work effectively according to its urgency and importance but this also requires self-discipline. You have to follow the schedule you have set for yourself!

+ Managing time and focusing on the task at hand means you will do better work and be less stressed, because you are not having to react to problems or crises. You can also find the time to include regular breaks and leisure activities in your schedule.

ACTION POINTS

✓ Find out how many assignments you can expect to receive this term and when you can expect to receive these. Enter this information into your student diary or onto a planner you can refer to regularly.

✓ Update your diary and/or planner with other commitments that you have this term – both work/college-related and social. Identify any potential clashes and decide the best action to take to solve the problem.

✓ Identify your own best time and place to work quietly and effectively.

✓ Displacement activities are things we do to put off starting a job we don't want to do – such as sending texts, watching TV, checking emails etc. Identify yours so that you know when you're doing them!

STEP FOUR

UTILISE ALL YOUR RESOURCES

Your resources are all the things that can help you to achieve your qualification. They can therefore be as wide-ranging as your favourite website and your **study buddy** (see page 15) who collects handouts for you if you miss a class.

Your college will provide the essential resources for your course, such as a library with a wide range of books and electronic reference sources, learning resource centre(s), the computer network and Internet access. Other basic resources you will be expected to provide yourself, such as file folders and paper. The policy on textbooks varies from one college to another, but on most courses today students are expected to buy their own. If you look after yours carefully, then you have the option to sell it on to someone else afterwards and recoup some of your money. If you scribble all over it, leave it on the floor and then tread on it, turn back pages and rapidly turn it into a dog-eared, misshapen version of its former self then you miss out on this opportunity.

Unfortunately students often squander other opportunities to utilise resources in the best way – usually because they don't think about them very much, if at all. To help, below is a list of the resources you should consider important – with a few tips on how to get the best out of them.

- **Course information** This includes your course specification, this Study Guide and all the other information relating to your BTEC National which you can find on the Edexcel website. Add to this all the information given to you at college relating to your course, including term dates, assignment dates and, of course, your timetable. This should not be 'dead' information that you glance at once and then discard or ignore. Rather it is important reference material that you need to store somewhere obvious, so that you can look at it whenever you have a query or need to clarify something quickly.

- **Course materials** In this group is your textbook (if there is one), the handouts you are given as well as print-outs and notes you make yourself. File handouts the moment you are given them and put them into an A4 folder bought for the purpose. You will need one for each unit you study. Some students prefer lever-arch files but these are more bulky so more difficult to carry around all day. Unless you have a locker at college it can be easier to keep a lever arch file at home for permanent storage of past handouts and notes for a unit and carry an A4 folder with you which contains current topic information. Filing handouts and print-outs promptly means they don't get lost. They are also less likely to get crumpled, torn or tatty becoming virtually unreadable. Unless you have a private and extensive source of income then this is even more important if you have to pay for every print-out you take in your college resource centre. If you are following a course such as Art and Design, then there will be all your art materials and the pieces you produce. You must look after these with great care.

- **Other stationery items** Having your own pens, pencils, notepad, punch, stapler and sets of dividers is essential. Nothing irritates tutors more than watching one punch circulate around a group – except, perhaps, the student who trudges into class with nothing to write on or with. Your dividers should be clearly labelled to help you store and find notes, print-outs and handouts fast. Similarly, your notes should be clearly headed and dated. If you are writing notes up from your own research then you will have to include your source. Researching information is explained in Step 6 – Sharpen your skills.

- **Equipment and facilities** These include your college library and resource centres, the college computer network and other college equipment you can use, such as laptop computers, photocopiers and presentation equipment. Much of this may be freely available; others – such as using the photocopier in the college library or the printers in a resource centre – may cost you money. Many useful resources will be electronic, such as DVDs or electronic journals and databases. At home you may have your own computer with Internet access to count as a resource. Finally, include any specialist equipment and facilities available for your particular course that you use at college or have at home.

Utilise all your resources

14

All centralised college resources and facilities are invaluable if you know how to use them – but can be baffling when you don't. Your induction should have included how to use the library, resource centre(s) and computer network. You should also have been informed of the policy on using IT equipment which determines what you can and can't do when you are using college computers. If, by any chance, you missed this then go and check it out for yourself. Library and resource centre staff will be only too pleased to give you helpful advice – especially if you pick a quiet time to call in. You can also find out about the allowable ways to transfer data between your college computer and your home computer if your options are limited because of IT security.

Having a study buddy is a good idea

- **People** You are surrounded by people who are valuable resources: your tutor(s), specialist staff at college, your employer and work colleagues, your relatives and any friends who have particular skills or who work in the same area you are studying. Other members of your class are also useful resources – although they may not always seem like it! Use them, for example, to discuss topics out of class time. A good debate between a group of students can often raise and clarify issues that there may not be time to discuss fully in class. Having a study buddy is another good idea – you get/make notes for them when they are away and vice versa. That way you don't miss anything.

  If you want information or help from someone, especially anyone outside your immediate circle, then remember to get the basics right! Approach them courteously, do your homework first so that you are well-prepared and remember that you are asking for assistance – not trying to get them to do the work for you! If someone has agreed to allow you to interview them as part of your research for an assignment or project then good preparations will be vital, as you will see in Step 6 – Sharpen your Skills (see page 22).

  One word of warning: be wary about using information from friends or relatives who have done a similar or earlier course. First, the slant of the material they were given may be different. It may also be out-of-date. And *never* copy anyone else's written assignments. This is **plagiarism** – a deadly sin in the educational world. You can read more about this in Step 5 – Understand your assessment (see page 16).

- **You!** You have the ability to be your own best resource or your own worst enemy! The difference depends upon your work skills, your personal skills and your attitude to your course and other people. You have already seen how to use time wisely. Throughout this guide you will find out how to sharpen and improve other work and personal skills and how to get the most out of your course – but it is up to you to read it and apply your new-found knowledge! This is why attributes like a positive attitude, an enquiring mind and the ability to focus on what is important all have a major impact on your final result.

PLUSPOINTS

+ Resources help you to achieve your qualification. You will squander these unwittingly if you don't know what they are or how to use them properly.

+ Course information needs to be stored safely for future reference: course materials need to be filed promptly and accurately so that you can find them quickly.

+ You need your own set of key stationery items; you also need to know how to use any central facilities or resources such as the library, learning resource centres and your computer network.

+ People are often a key resource – school or college staff, work colleagues, members of your class, people who are experts in their field.

+ You can be your own best resource! Develop the skills you need to be able to work quickly and accurately and to get the most out of other people who can help you.

ACTION POINTS

✓ Under the same headings as in this section, list all the resources you need for your course and tick off those you currently have. Then decide how and when you can obtain anything vital that you lack.

✓ Check that you know how to access and use all the shared resources to which you have access at school or college.

✓ Pair up with someone on your course as a study buddy – and don't let them down!

✓ Test your own storage systems. How fast can you find notes or print-outs you made yesterday/last week/last month – and what condition are they in?

✓ Find out the IT policy at your school or college and make sure you abide by it.

16

STEP FIVE

UNDERSTAND YOUR ASSESSMENT

The key to doing really, really well on any BTEC National course is to understand exactly what you are expected to do in your assignments – and then to do it! It really is as simple as that. So why is it that some people go wrong?

Obviously you may worry that an assignment may be so difficult that it is beyond you. Actually this is highly unlikely to happen because all your assignments are based on topics you will have already covered thoroughly in class. Therefore, if you have attended regularly – and have clarified any queries or worries you have either in class or during your tutorials – this shouldn't happen. If you have had an unavoidable lengthy absence then you may need to review your progress with your tutor and decide how best to cope with the situation. Otherwise, you should note that the main problems with assignments are usually due to far more mundane pitfalls – such as:

☐ not reading the instructions or the assignment brief properly

☐ not understanding what you are supposed to do

☐ only doing part of the task or answering part of a question

☐ skimping the preparation, the research or the whole thing

☐ not communicating your ideas clearly

☐ guessing answers rather than researching properly

☐ padding out answers with irrelevant information

☐ leaving the work until the last minute and then doing it in a rush

☐ ignoring advice and feedback your tutor has given you.

You can avoid all of these traps by following the guidelines below so that you know exactly what you are doing, prepare well and produce your best work.

The assignment 'brief'

The word 'brief' is just another way of saying 'instructions'. Often, though, a 'brief' (despite its name!) may be rather longer. The brief sets the context for the work, defines what evidence you will need to produce and matches the grading criteria to the tasks. It will also give you a schedule for completing the tasks. For example, a brief may include details of a case study you have to read; research you have to carry out or a task you have to do, as well as questions you have to answer. Or it may give you details about a project or group presentation you have to prepare. The type of assignments you receive will depend partly upon the vocational area you are studying, but you can expect some to be in the form of written assignments. Others are likely to be more practical or project-based, especially if you are doing a very practical subject such as Art and Design, Performing Arts or Sport. You may also be assessed in the workplace. For example, this is a course requirement if you are studying Children's Care, Learning and Development.

The assignment brief may also include the **learning outcomes** to which it relates. These tell you the purpose of the assessment and the knowledge you need to demonstrate to obtain a required grade. If your brief doesn't list the learning outcomes, then you should check this information against the unit specification to see the exact knowledge you need to demonstrate.

The grade(s) you can obtain will also be stated on the assignment brief. Sometimes an assignment will focus on just one grade. Others may give you the opportunity to develop or extend your work to progress to a higher grade. This is often dependent upon submitting acceptable work at the previous grade first. You will see examples of this in the Marked Assignments section of this Study Guide on page 157.

The brief will also tell you if you have to do part of the work as a member of a group. In this case, you must identify your own contribution. You may also be expected to take part in a **peer review**, where you all give feedback on the contribution of one another. Remember that you should do this as objectively and professionally as possible – not just praise everyone madly in the hope that they will do the same for you! In any assignment where there is a group contribution, there is virtually always an individual component, so that your individual grade can be assessed accurately.

Finally, your assignment brief should state the final deadline for handing in the work as well as any interim review dates when you can discuss your progress and ideas with your tutor. These are very important dates indeed and should be entered immediately into your diary or planner. You should schedule your work around these dates so that you have made a start by the first date. This will then enable you to note any queries or significant issues you want to discuss. Otherwise you will waste a valuable opportunity to obtain useful feedback on your progress. Remember, too, to take a notebook to any review meetings so that you can write down the guidance you are given.

Your school or college rules and regulations

Your school or college will have a number of policies and guidelines about assignments and assessment. These will deal with issues such as:

- The procedure you must follow if you have a serious personal problem so cannot meet the deadline date and need an extension.
- Any penalties for missing a deadline date without any good reason.
- The penalties for copying someone else's work (**plagiarism**). These will be severe so make sure that you never share your work (including your CDs) with anyone else and don't ask to borrow theirs.
- The procedure to follow if you are unhappy with the final grade you receive.

Even though it is unlikely that you will ever need to use any of these policies, it is sensible to know they exist, and what they say, just as a safeguard.

Understanding the question or task

There are two aspects to a question or task that need attention. The first are the *command words*, which are explained below. The second are the *presentation instructions*, so that if you are asked to produce a table or graph or report then you do exactly that – and don't write a list or an essay instead!

Command words are used to specify how a question must be answered, eg 'explain', 'describe', 'analyse', 'evaluate'. These words relate to the type of answer required. So whereas you may be asked to 'describe' something at Pass level, you will need to do more (such as 'analyse' or 'evaluate') to achieve Merit or Distinction grade.

Many students fail to get a higher grade because they do not realise the difference between these words. They simply don't know *how* to analyse or evaluate, so give an explanation instead. Just adding to a list or giving a few more details will never give you a higher grade – instead you need to change your whole approach to the answer.

The **grading grid** for each unit of your course gives you the command words, so that you can find out exactly what you have to do in each unit, to obtain a Pass, Merit and Distinction. The following charts show you what is usually required when you see a particular command word. You can use this, and the marked assignments on pages 157–175, to see the difference between the types of answers required for each grade. (The assignments your centre gives you will be specially written to ensure you have the opportunity to achieve all the possible grades.) Remember, though, that these are just examples to guide you. The exact response will often depend upon the way a question is worded, so if you have any doubts at all check with your tutor before you start work.

There are two other important points to note:

- Sometimes the same command word may be repeated for different grades – such as 'create' or 'explain'. In this case the *complexity* or *range* of the task itself increases at the higher grades – as you will see if you read the grading grid for the unit.

- Command words can also vary depending upon your vocational area. If you are studying Performing Arts or Art and Design you will probably find several command words that an Engineer or IT Practitioner would not – and vice versa!

To obtain a Pass grade

To achieve this grade you must usually demonstrate that you understand the important facts relating to a topic and can state these clearly and concisely.

Command word	What this means
Create (or produce)	Make, invent or construct an item.
Describe	Give a clear, straightforward description that includes all the main points and links these together logically.
Define	Clearly explain what a particular term means and give an example, if appropriate, to show what you mean.
Explain . . . how/why	Set out in detail the meaning of something, with reasons. It is often helpful to give an example of what you mean. Start with the topic then give the 'how' or 'why'.
Identify	Distinguish and state the main features or basic facts relating to a topic.
Interpret	Define or explain the meaning of something.
Illustrate	Give examples to show what you mean.
List	Provide the information required in a list rather than in continuous writing.
Outline	Write a clear description that includes all the main points but avoid going into too much detail.
Plan (or devise)	Work out and explain how you would carry out a task or activity.
Select (and present) information	Identify relevant information to support the argument you are making and communicate this in an appropriate way.
State	Write a clear and full account.
Undertake	Carry out a specific activity.
Examples: **Identify** the main features on a digital camera. **Describe** your usual lifestyle. **Outline** the steps to take to carry out research for an assignment.	

To obtain a Merit grade

To obtain this grade you must prove that you can apply your knowledge in a specific way.

Command word	What this means
Analyse	Identify separate factors, say how they are related and how each one relates to the topic.
Classify	Sort your information into appropriate categories before presenting or explaining it.
Compare and contrast	Identify the main factors that apply in two or more situations and explain the similarities and differences or advantages and disadvantages.
Demonstrate	Provide several relevant examples or appropriate evidence which support the arguments you are making. In some vocational areas this may also mean giving a practical performance.
Discuss	Provide a thoughtful and logical argument to support the case you are making.
Explain (in detail)	Provide details and give reasons and/or evidence to clearly support the argument you are making.
Implement	Put into practice or operation. You may also have to interpret or justify the effect or result.
Interpret	Understand and explain an effect or result.
Justify	Give appropriate reasons to support your opinion or views and show how you arrived at these conclusions.
Relate/report	Give a full account of, with reasons.
Research	Carry out a full investigation.
Specify	Provide full details and descriptions of selected items or activities.
Examples: **Compare and contrast** the performance of two different digital cameras. **Justify** your usual lifestyle. **Explain in detail** the steps to take to research an assignment.	

To obtain a Distinction grade

To obtain this grade you must prove that you can make a reasoned judgement based on appropriate evidence.

Command word	What this means
Analyse	Identify the key factors, show how they are linked and explain the importance and relevance of each.
Assess	Give careful consideration to all the factors or events that apply and identify which are the most important and relevant with reasons for your views.
Comprehensively explain	Give a very detailed explanation that covers all the relevant points and give reasons for your views or actions.
Comment critically	Give your view after you have considered all the evidence, particularly the importance of both the relevant positive and negative aspects.
Evaluate	Review the information and then bring it together to form a conclusion. Give evidence to support each of your views or statements.
Evaluate critically	Review the information to decide the degree to which something is true, important or valuable. Then assess possible alternatives taking into account their strengths and weaknesses if they were applied instead. Then give a precise and detailed account to explain your opinion.
Summarise	Identify review the main, relevant factors and/or arguments so that these are explained in a clear and concise manner.
Examples: **Assess** ten features commonly found on a digital camera. **Evaluate critically** your usual lifestyle. **Analyse** your own ability to carry out effective research for an assignment.	

Responding positively

This is often the most important attribute of all! If you believe that assignments give you the opportunity to demonstrate what you know and how you can apply it *and* positively respond to the challenge by being determined to give it your best shot, then you will do far better than someone who is defeated before they start.

It obviously helps, too, if you are well organised and have confidence in your own abilities – which is what the next section is all about!

PLUSPOINTS

+ Many mistakes in assignments are through errors that can easily be avoided such as not reading the instructions properly or doing only part of the task that was set!

+ Always read the assignment brief very carefully indeed. Check that you understand exactly what you have to do and the learning outcomes you must demonstrate.

+ Make a note of the deadline for an assignment and any interim review dates on your planner. Schedule work around these dates so that you can make the most of reviews with your tutor.

+ Make sure you know about school or college policies relating to assessment, such as how to obtain an extension or query a final grade.

+ For every assignment, make sure you understand the command words, which tell you how to answer the question, and the presentation instructions, which say what you must produce.

+ Command words are shown in the grading grid for each unit of your qualification. Expect command words and/or the complexity of a task to be different at higher grades, because you have to demonstrate higher-level skills.

ACTION POINTS

✓ Discuss with your tutor the format (style) of assignments you are likely to receive on your course, eg assignments, projects, or practical work where you are observed.

✓ Check the format of the assignments in the Marked Assignments section of this book. Look at the type of work students did to gain a Pass and then look at the difference in the Merit answers. Read the tutor's comments carefully and ask your own tutor if there is anything you do not understand.

✓ Check out all the policies and guidelines at your school or college that relate to assessment and make sure you understand them.

✓ Check out the grading grid for the units you are currently studying and identify the command words for each grade. Then check you understand what they mean using the explanations above. If there are any words that are not included, ask your tutor to explain the meanings and what you would be required to do.

STEP SIX

SHARPEN YOUR SKILLS

To do your best in any assignment you need a number of skills. Some of these may be vocationally specific, or professional, skills that you are learning as part of your course – such as acting or dancing if you are taking a Performing Arts course or, perhaps, football if you are following a Sports course. Others, though, are broader skills that will help you to do well in assignments no matter what subjects or topics you are studying – such as communicating clearly and cooperating with others.

Some of these skills you will have already and in some areas you may be extremely proficient. Knowing where your weaknesses lie, though, and doing something about them has many benefits. You will work more quickly, more accurately *and* have increased confidence in your own abilities. As an extra bonus, all these skills also make you more effective at work – so there really is no excuse for not giving yourself a quick skills check and then remedying any problem areas.

This section contains hints and tips to help you check out and improve each of the following areas:

- Your numeracy skills
- Keyboarding and document preparation
- Your IT skills
- Your written communication skills
- Working with others
- Researching information
- Making a presentation
- Problem solving and staying focused

Improving your numeracy skills

Some people have the idea that they can ignore numeracy because this skill isn't relevant to their vocational area – such as Art and Design or Children's Care, Learning and Development. If this is how you think then you are wrong! Numeracy is a life skill that everyone needs, so if you can't carry out basic calculations accurately then you will have problems, often when you least expect them.

Fortunately there are several things you can do to remedy this situation:

- Practise basic calculations in your head and then check them on a calculator.
- Ask your tutor if there are any essential calculations which give you difficulties.
- Use your onscreen calculator (or a spreadsheet package) to do calculations for you when you are using your computer.
- Try your hand at Sudoku puzzles – either on paper or by using a software package or online at sites such as www.websudoku.com/.
- Investigate puzzle sites and brain training software, such as http://school.discovery.com/brainboosters/ and Dr Kawashima's Brain Training by Nintendo.
- Check out online sites such as www.bbc.co.uk/skillswise/ and www.bbc.co.uk/schools/ks3bitesize/maths/number/index.shtml to improve your skills.

Numeracy is a life skill

Keyboarding and document preparation

- Think seriously about learning to touch type to save hours of time! Your school or college may have a workshop you can join or you can learn online such as by downloading a free program at www.sense-lang. org/typing/ or practising on sites such as www.computerlab.kids.new. net/keyboarding.htm.
- Obtain correct examples of document formats you will have to use, such as a report or summary. Your tutor may provide you with these or you can find examples in many communication textbooks.
- Proof-read work you produce on a computer *carefully*. Remember that your spell checker will not pick up every mistake you make, such as a mistyped word that makes another word (eg form/from; sheer/shear)

and grammar checkers, too, are not without their problems! This means you still have to read your work through yourself. If possible, let your work go 'cold' before you do this so that you read it afresh and don't make assumptions about what you have written. Then read word by word to make sure it still makes sense and there are no silly mistakes, such as missing or duplicated words.

- Make sure your work looks professional by using an appropriate typeface and font size as well as suitable margins.
- Print out your work carefully and store it neatly, so it looks in pristine condition when you hand it in.

Your IT skills

- Check that you can use the main features of all the software packages that you will need to produce your assignments, such as Word, Excel and PowerPoint.
- Adopt a good search engine, such as Google, and learn to use it properly. Many have online tutorials such as www.googleguide.com.
- Develop your IT skills to enable you to enhance your assignments appropriately. For example, this may include learning how to import and export text and artwork from one package to another; taking digital photographs and inserting them into your work and/or creating drawings or diagrams by using appropriate software for your course.

Your written communication skills

A poor vocabulary will reduce your ability to explain yourself clearly; work peppered with spelling or punctuation errors looks unprofessional.

- Read more. This introduces you to new words and familiarises you over and over again with the correct way to spell words.
- Look up words you don't understand in a dictionary and then try to use them yourself in conversation.
- Use the Thesaurus in Word to find alternatives to words you find yourself regularly repeating, to add variety to your work.
- *Never* use words you don't understand in the hope that they sound impressive!
- Do crosswords to improve your word power and spelling.
- Resolve to master punctuation – especially apostrophes – either by using an online programme or working your way through the relevant section of a communication textbook that you like.
- Check out online sites such as www.bbc.co.uk/skillswise/ and www.bbc.co.uk/schools/gcsebitesize/english/ as well as puzzle sites with communication questions such as http://school.discovery.com/brainboosters/.

Working with others

In your private life you can choose who you want to be with and how you respond to them. At work you cannot do that – you are paid to be professional and this means working alongside a wide variety of people, some of whom you may like and some of whom you may not!

The same applies at school or college. By the time you have reached BTEC National level you will be expected to have outgrown wanting to work with your best friends on every project! You may not be very keen on everyone who is in the same team as you, but – at the very least – you can be pleasant, cooperative and helpful. In a large group this isn't normally too difficult. You may find it much harder if you have to partner someone who has very different ideas and ways of working to you.

In this case it may help if you:

- Realise that everyone is different and that your ways of working may not always be the best!
- Are prepared to listen and contribute to a discussion (positively) in equal amounts. Make sure, too, that you encourage the quiet members of the group to speak up by asking them what their views are. The ability to draw other people into the discussion is an important and valuable skill to learn.
- Write down what you have said you will do, so that you don't forget anything.
- Are prepared to do your fair share of the work.
- Discuss options and alternatives with people – don't give them orders or meekly accept instructions and then resent it afterwards.
- Don't expect other people to do what you wouldn't be prepared to do.
- Are sensitive to other people's feelings and remember that they may have personal problems or issues that affect their behaviour.
- *Always* keep your promises and never let anyone down when they are depending upon you.
- Don't flounce around or lose your temper if things get tough. Instead take a break while you cool down. Then sit down and discuss the issues that are annoying you.
- Help other people to reach a compromise when necessary, by acting as peacemaker.

Researching information

Poor researchers either cannot find what they want or find too much – and then drown in a pile of papers. If you find yourself drifting aimlessly around a library when you want information or printing out dozens of pages for no apparent purpose, then this section is for you!

- Always check *exactly* what it is you need to find and how much detail is needed. Write down a few key words to keep yourself focused.
- Discipline yourself to ignore anything that is irrelevant – from books with interesting titles to websites which sound tempting but have little to do with your topic or key words.
- Remember that you could theoretically research information forever! So at some time you have to call a halt. Learning when to do this is another skill, but you can learn this by writing out a schedule which clearly states when you have to stop looking and start sorting out your information and writing about it!

- In a library, check you know how the books are stored and what other types of media are available. If you can't find what you are looking for then ask the librarian for help. Checking the index in a book is the quickest way to find out whether it contains information related to your key words. Put it back if it doesn't or if you can't understand it. If you find three or four books and/or journals that contain what you need then that is usually enough.

- Online use a good search engine and use the summary of the search results to check out the best sites. Force yourself to check out sites beyond page one of the search results! When you enter a site investigate it carefully – use the site map if necessary. It isn't always easy to find exactly what you want. Bookmark sites you find helpful and will want to use again and only take print-outs when the information is closely related to your key words.

- Talk to people who can help you (see also Step 4 – Utilise all your resources) and prepare in advance by thinking about the best questions to ask. Always explain why you want the information and don't expect anyone to tell you anything that is confidential or sensitive – such as personal information or financial details. Always write clear notes so that you remember what you have been told, by whom and when. If you are wise you will also note down their contact details so that you can contact them again if you think of anything later. If you remember to be courteous and thank them for their help, this shouldn't be a problem.

- Store all your precious information carefully and neatly in a labelled folder so that you can find it easily. Then, when you are ready to start work, reread it and extract that which is most closely related to your key words and the task you are doing.

- Make sure you state the source of all the information you quote by including the name of the author or the web address, either in the text or as part of a bibliography at the end. Your school or college will have a help sheet which will tell you exactly how to do this.

Making a presentation

This involves several skills – which is why it is such a popular way of finding out what students can do! It will test your ability to work in a team, speak in public and use IT (normally PowerPoint) – as well as your nerves. It is therefore excellent practice for many of the tasks you will have to do when you are at work – from attending an interview to talking to an important client.

You will be less nervous if you have prepared well and have rehearsed your role beforehand. You will produce a better, more professional presentation if you take note of the following points.

- If you are working as a team, work out everyone's strengths and weaknesses and divide up the work (fairly) taking these into account. Work out, too, how long each person should speak and who would be the best as the 'leader' who introduces each person and then summarises everything at the end.

PLUSPOINTS

+ Poor numeracy skills can let you down in your assignments and at work. Work at improving these if you regularly struggle with even simple calculations.

+ Good keyboarding, document production and IT skills can save you hours of time and mean that your work is of a far more professional standard. Improve any of these areas which are letting you down.

+ Your written communication skills will be tested in many assignments. Work at improving areas of weakness, such as spelling, punctuation or vocabulary.

+ You will be expected to work cooperatively with other people both at work and during many assignments. Be sensitive to other people's feelings, not just your own, and always be prepared to do your fair share of the work and help other people when you can.

+ To research effectively you need to know exactly what you are trying to find and where to look. This means understanding how reference media is stored in your library as well as how to search online. Good organisation skills also help so that you store important information carefully and can find it later. And never forget to include your sources in a bibliography.

+ Making a presentation requires several skills and may be nerve-racking at first. You will reduce your problems if you prepare well, are not too ambitious and have several run-throughs beforehand. Remember to speak clearly and a little more slowly than normal and smile from time to time!

ACTION POINTS

✓ Test both your numeracy and literacy skills at http://www.move-on.org.uk/testyourskills.asp# to check your current level. You don't need to register on the site if you click to do the 'mini-test' instead. If either need improvement, get help at http://www.bbc.co.uk/keyskills/it/1.shtml.

✓ Do the following two tasks with a partner to jerk your brain into action!

  – Each write down 36 simple calculations in a list, eg 8 x 6, 19 – 8, 14 + 6. Then exchange lists. See who can answer the most correctly in the shortest time.

  – Each write down 30 short random words (no more than 8 letters), eg cave, table, happily. Exchange lists. You each have three minutes to try to remember as many words as possible. Then hand back the list and write down all those you can recall. See who can remember the most.

✓ Assess your own keyboarding, proof-reading, document production, written communication and IT skills. Then find out if your tutors agree with you!

✓ List ten traits in other people that drive you mad. Then, for each one, suggest what you could do to cope with the problem (or solve it) rather than make a fuss. Compare your ideas with other members of your group.

✓ Take a note of all feedback you receive from your tutors, especially in relation to working with other people, researching and giving presentations. In each case focus on their suggestions and ideas so that you continually improve your skills throughout the course.

- Don't be over-ambitious. Take account of your time-scale, resources and the skills of the team. Remember that a simple, clear presentation is often more professional than an over-elaborate or complicated one where half the visual aids don't work properly!

- If you are using PowerPoint try to avoid preparing every slide with bullet points! For variety, include some artwork and vary the designs. Remember that you should *never* just read your slides to the audience! Instead prepare notes that you can print out that will enable you to enhance and extend what the audience is reading.

- Your preparations should also include checking the venue and time; deciding what to wear and getting it ready; preparing, checking and printing any handouts; deciding what questions might be asked and how to answer these.

- Have several run-throughs beforehand and check your timings. Check, too, that you can be heard clearly. This means lifting up your head and 'speaking' to the back of the room a little more slowly and loudly than you normally do.

- On the day, arrive in plenty of time so that you aren't rushed or stressed. Remember that taking deep breaths helps to calm your nerves.

- Start by introducing yourself clearly and smile at the audience. If it helps, find a friendly face and pretend you are just talking to that person.

- Answer any questions honestly and don't exaggerate, guess or waffle. If you don't know the answer then say so!

- If you are giving the presentation in a team, help out someone else who is struggling with a question if you know the answer.

- Don't get annoyed or upset if you get any negative feedback afterwards. Instead take note so that you can concentrate on improving your own performance next time. And don't focus on one or two criticisms and ignore all the praise you received! Building on the good and minimising the bad is how everyone improves in life!

STEP SEVEN

MAXIMISE YOUR OPPORTUNITIES AND MANAGE YOUR PROBLEMS

Like most things in life, you may have a few ups and downs on your course – particularly if you are studying over quite a long time, such as one or two years. Sometimes everything will be marvellous – you are enjoying all the units, you are up-to-date with your work, you are finding the subjects interesting and having no problems with any of your research tasks. At other times you may struggle a little more. You may find one or two topics rather tedious, or there may be distractions or worries in your personal life that you have to cope with. You may struggle to concentrate on the work and do your best.

Rather than just suffering in silence or gritting your teeth if things go a bit awry it is sensible if you have an action plan to help you cope. Equally, rather than just accepting good opportunities for additional experiences or learning, it is also wise to plan how to make the best of these. This section will show you how to do this.

Making the most of your opportunities

The following are examples of opportunities to find out more about information relevant to your course or to try putting some of your skills into practice.

- **External visits** You may go out of college on visits to different places or organisations. These are not days off – there is a reason for making each trip. Prepare in advance by reading around relevant topics and make notes of useful information whilst you are there. Then write (or type) it up neatly as soon as you can and file it where you can find it again!

- **Visiting speakers** Again, people are asked to talk to your group for a purpose. You are likely to be asked to contribute towards questions that may be asked – which may be submitted in advance so that the speaker is clear on the topics you are studying. Think carefully about information that you would find helpful so that you can ask one or two relevant and useful questions. Take notes whilst the speaker is addressing your group, unless someone is recording the session. Be prepared to thank the speaker on behalf of your group if you are asked to do so.

- **Professional contacts** These will be the people you meet on work experience doing the real job that one day you hope to do. Make the most of meeting these people to find out about the vocational area of your choice.

- **Work experience** If you need to undertake practical work for any particular units of your BTEC National qualification, and if you are studying full-time, then your tutor will organise a work experience placement for you and talk to you about the evidence you need to obtain. You may also be issued with a special log book or diary in which to record your experiences. Before you start your placement, check that you are clear about all the details, such as the time you will start and leave, the name of your supervisor, what you should wear and what you should do if you are ill during the placement and cannot attend. Read and reread the units to which your evidence will apply and make sure you understand the grading criteria and what you need to obtain. Then make a note of appropriate headings to record your information. Try to make time to write up your notes, log book and/or diary every night, whilst your experiences are fresh in your mind.

- **In your own workplace** You may be studying your BTEC National qualification on a part-time basis and also have a full-time job in the same vocational area. Or you may be studying full-time and have a part-time job just to earn some money. In either case you should be alert to opportunities to find out more about topics that relate to your workplace, no matter how generally. For example, many BTEC courses include topics such as health and safety, teamwork, dealing with customers, IT security and communications – to name but a few. All these are topics that your employer will have had to address and finding out more about these will broaden your knowledge and help to give more depth to your assignment responses.

- **Television programmes, newspapers, Podcasts and other information sources.** No matter what vocational area you are studying, the media are likely to be an invaluable source of information. You should be alert to any news bulletins that relate to your studies as well as relevant information in more topical television programmes. For example, if you are studying Art and Design then you should make a particular effort to watch the *Culture Show* as well as programmes on artists, exhibitions or other topics of interest. Business students should find inspiration by watching *Dragons Den*, *The Apprentice* and the *Money Programme* and Travel and Tourism students should watch holiday, travel and adventure programmes. If you are studying Media, Music and Performing Arts then you are spoiled for choice! Whatever your vocational choice, there will be television and radio programmes of special interest to you.

Remember that you can record television programmes to watch later if you prefer, and check out newspaper headlines online and from sites such as BBC news. The same applies to Podcasts. Of course, to know which information is relevant means that you must be familiar with the content of all the units you are studying, so it is useful to know what topics you will be learning about in the months to come, as well as the ones you are covering now. That way you can recognise useful opportunities when they arise.

The media are invaluable sources of information

Minimising problems

If you are fortunate, any problems you experience on your course will only be minor ones. For example, you may struggle to keep yourself motivated every single day and there may be times that you are having difficulty with a topic. Or you may be struggling to work with someone else in your team or to understand a particular tutor.

During induction you should have been told which tutor to talk to in this situation, and who to see if that person is absent or if you would prefer to see someone else. If you are having difficulties which are distracting you and affecting your work then it is sensible to ask to see your tutor promptly so that you can talk in confidence, rather than just trusting to luck everything will go right again. It is a rare student who is madly enthusiastic about every part of a course and all the other people on the course, so your tutor won't be surprised and will be able to give you useful guidance to help you stay on track.

If you are very unlucky, you may have a more serious personal problem to deal with. In this case it is important that you know the main sources of help in your school or college and how to access these.

■ **Professional counselling** There may be a professional counselling service if you have a concern that you don't want to discuss with any teaching staff. If you book an appointment to see a counsellor then you can be certain that nothing you say will ever be mentioned to another member of staff without your permission.

■ **Student complaint procedures** If you have a serious complaint to make then the first step is to talk to a tutor, but you should be aware of the formal student complaint procedures that exist if you cannot resolve the problem informally. Note that these are only used for serious issues, not for minor difficulties.

■ **Student appeals procedures** If you cannot agree with a tutor about a final grade for an assignment then you need to check the grading criteria and ask the tutor to explain how the grade was awarded. If you are still unhappy then you should see your personal tutor. If you still disagree then you have the right to make a formal appeal.

- **Student disciplinary procedures** These exist so that all students who flout the rules in a school or college will be dealt with in the same way. Obviously it is wise to avoid getting into trouble at any time, but if you find yourself on the wrong side of the regulations do read the procedures carefully to see what could happen. Remember that being honest about what happened and making a swift apology is always the wisest course of action, rather than being devious or trying to blame someone else.

- **Serious illness** Whether this affects you or a close family member, it could severely affect your attendance. The sooner you discuss the problem with your tutor the better. This is because you will be missing notes and information from the first day you do not attend. Many students under-estimate the ability of their tutors to find inventive solutions in this type of situation – from sending notes by post to updating you electronically if you are well enough to cope with the work.

PLUSPOINTS

+ Some students miss out on opportunities to learn more about relevant topics. This may be because they haven't read the unit specifications, so don't know what topics they will be learning about in future; haven't prepared in advance or don't take advantage of occasions when they can listen to an expert and perhaps ask questions. Examples of these occasions include external visits, visiting speakers, work experience, being at work and watching television.

+ Many students encounter minor difficulties, especially if their course lasts a year or two. It is important to talk to your tutor, or another appropriate person, promptly if you have a worry that is affecting your work.

+ All schools and colleges have procedures for dealing with important issues and problems such as serious complaints, major illnesses, student appeals and disciplinary matters. It is important to know what these are.

ACTION POINTS

✓ List the type of opportunities available on your course for obtaining more information and talking to experts. Then check with your tutor to make sure you haven't missed out any.

✓ Check out the content of each unit you will be studying so that you know the main topics you have still to study.

✓ Identify the type of information you can find on television, in newspapers and in Podcasts that will be relevant to your studies.

✓ Check out your school or college documents and procedures to make sure that you know who to talk to in a crisis and who you can see if the first person is absent.

✓ Find out where you can read a copy of the main procedures in your school or college that might affect you if you have a serious problem. Then do so.

AND FINALLY . . .

Don't expect this Introduction to be of much use if you skim through it quickly and then put it to one side. Instead, refer to it whenever you need to remind yourself about something related to your course.

The same applies to the rest of this Student Guide. The Activities in the next section have been written to help you to demonstrate your understanding of many of the key topics contained in the core or specialist units you are studying. Your tutor may tell you to do these at certain times; otherwise there is nothing to stop you working through them yourself!

Similarly, the Marked Assignments in the final section have been written to show you how your assignments may be worded. You can also see the type of response that will achieve a Pass, Merit and Distinction – as well as the type of response that won't! Read these carefully and if any comment or grade puzzles you, ask your tutor to explain it.

Then keep this guide in a safe place so that you can use it whenever you need to refresh your memory. That way, you will get the very best out of your course – and yourself!

GLOSSARY

Note: all words highlighted in bold in the text are defined in the glossary.

Accreditation of Prior Learning (APL)

APL is an assessment process that enables your previous achievements and experiences to count towards your qualification providing your evidence is authentic, current, relevant and sufficient.

Apprenticeships

Schemes that enable you to work and earn money at the same time as you gain further qualifications (an **NVQ** award and a technical certificate) and improve your key skills. Apprentices learn work-based skills relevant to their job role and their chosen industry. You can find out more at www.apprenticeships.org.uk/

Assessment methods

Methods, such as **assignments**, case studies and practical tasks, used to check that your work demonstrates the learning and understanding required for your qualification.

Assessor

The tutor who marks or assesses your work.

Assignment

A complex task or mini-project set to meet specific **grading criteria**.

Awarding body

The organisation which is responsible for devising, assessing and issuing qualifications. The awarding body for all BTEC qualifications is Edexcel.

Core units

On a BTEC National course these are the compulsory or mandatory units that all students must complete to gain the qualification. Some BTEC qualifications have an over-arching title, eg Engineering, but within Engineering you can choose different routes. In this case you will study both common core units that are common to all engineering qualifications and **specialist core unit(s)** which are specific to your chosen **pathway**.

DCSF

The Department for Children, Schools and Families: this is the government department responsible for education issues. You can find out more at www.dcsf.gov.uk

32

Degrees

These are higher education qualifications which are offered by universities and colleges. Foundation degrees take two years to complete; honours degrees may take three years or longer. See also **Higher National Certificates and Diplomas**.

Distance learning

This enables you to learn and/or study for a qualification without attending an Edexcel centre although you would normally be supported by a member of staff who works there. You communicate with your tutor and/or the centre that organises the distance learning programme by post, telephone or electronically.

Educational Maintenance Award (EMA)

This is a means-tested award which provides eligible students under 19, who are studying a full-time course at school or college, with a cash sum of money every week. See http://www.ema.direct.gov.uk for up-to-date details.

External verification

Formal checking by a representative of Edexcel of the way a BTEC course is delivered. This includes sampling various assessments to check content and grading.

Final major project

This is a major, individual piece of work that is designed to enable you to demonstrate you have achieved several learning outcomes for a BTEC National qualification in the creative or performing arts. Like all assessments, this is internally assessed.

Forbidden combinations

Qualifications or units that cannot be taken simultaneously because their content is too similar.

GLH

See **Guided Learning Hours** on page 34.

Grade

The rating (Pass, Merit or Distinction) given to the mark you have obtained which identifies the standard you have achieved.

Grade boundaries

The pre-set points at which the total points you have earned for different units converts to the overall grade(s) for your qualification.

Grading criteria

The standard you have to demonstrate to obtain a particular grade in the unit, in other words, what you have to prove you can do.

Grading domains

The main areas of learning which support the **learning outcomes**. On a BTEC National course these are: application of knowledge and understanding; development of practical and technical skills; personal development for occupational roles; application of generic and **key skills**. Generic skills are basic skills needed wherever you work, such as the ability to work cooperatively as a member of a team.

Grading grid

The table in each unit of your BTEC qualification specification that sets out the **grading criteria**.

Guided Learning Hours (GLH)

The approximate time taken to deliver a unit which includes the time taken for direct teaching, instruction and assessment and for you to carry out directed assignments or directed individual study. It does not include any time you spend on private study or researching an assignment. The GLH determines the size of the unit. At BTEC National level, units are either 30, 60, 90 or 120 guided learning hours. By looking at the number of GLH a unit takes, you can see the size of the unit and how long it is likely to take you to learn and understand the topics it contains.

Higher education (HE)

Post-secondary and post-further education, usually provided by universities and colleges.

Higher level skills

Skills such as evaluating or critically assessing complex information that are more difficult than lower level skills such as writing a description or making out a list. You must be able to demonstrate higher level skills to achieve a Distinction grade.

Higher National Certificates and Diplomas

Higher National Certificates and Diplomas are vocational qualifications offered at colleges around the country. Certificates are part-time and designed to be studied by people who are already in work; students can use their work experiences to build on their learning. Diplomas are full-time courses – although often students will spend a whole year on work experience part way through their Diploma. Higher Nationals are roughly equivalent to half a degree.

Indicative reading

Recommended books and journals whose content is both suitable and relevant for the unit.

Induction

A short programme of events at the start of a course designed to give you essential information and introduce you to your fellow students and tutors so that you can settle down as quickly and easily as possible.

Internal verification

The quality checks carried out by nominated tutor(s) at your school or college to ensure that all assignments are at the right level and cover appropriate learning outcomes. The checks also ensure that all **assessors** are marking work consistently and to the same standard.

Investors in People (IIP)

A national quality standard which sets a level of good practice for the training and development of people. Organisations must demonstrate their commitment to achieve the standard.

Key skills

The transferable, essential skills you need both at work and to run your own life successfully. They are: literacy, numeracy, ICT, problem solving, working with others and self-management.

Learning outcomes

The knowledge and skills you must demonstrate to show that you have effectively learned a unit.

Learning support

Additional help that is available to all students in a school or college who have learning difficulties or other special needs. These include reasonable adjustments to help to reduce the effect of a disability or difficulty that would place a student at a substantial disadvantage in an assessment situation.

Levels of study

The depth, breadth and complexity of knowledge, understanding and skills required to achieve a qualification determines its level. Level 2 is broadly equivalent to GCSE level (grades A*-C) and level 3 equates to GCE level. As you successfully achieve one level, you can then progress on to the next. BTEC qualifications are offered at Entry level, then levels 1, 2, 3, 4 and 5.

Learning and Skills Council (LSC)

The government body responsible for planning and funding education and training for everyone aged over 16 in England – except university students. You can find out more at www.lsc.gov.uk

Local Education Authority (LEA)

The local government body responsible for providing education for students of compulsory school age in your area.

Mentor

A more experienced person who will guide and counsel you if you have a problem or difficulty.

Mode of delivery

The way in which a qualification is offered to students, eg part-time, full-time, as a short course or by **distance learning**.

National Occupational Standard (NOS)

These are statements of the skills, knowledge and understanding you need to develop to be competent at a particular job. These are drawn up by the **Sector Skills Councils**.

National Qualification Framework (NQF)

The framework into which all accredited qualifications in the UK are placed. Each is awarded a level based on their difficulty which ensures that all those at the same level are of the same standard. (See also **levels of study**.)

National Vocational Qualification (NVQ)

Qualifications which concentrate upon the practical skills and knowledge required to do a job competently. They are usually assessed in the workplace and range from level 1 (the lowest) to level 5 (the highest).

Nested qualifications

Qualifications which have 'common' units, so that students can easily progress from one to another by adding on more units, such as the BTEC Award, BTEC Certificate and BTEC Diploma.

Pathway

All BTEC National qualifications are comprised of a small number of core units and a larger number of specialist units. These specialist units are grouped into different combinations to provide alternative pathways to achieving the qualification, linked to different career preferences.

Peer review

An occasion when you give feedback on the performance of other members in your team and they, in turn, comment on your performance.

Plagiarism

The practice of copying someone else's work and passing it off as your own. *This is strictly forbidden on all courses.*

Portfolio

A collection of work compiled by a student, usually as evidence of learning to produce for an **assessor**.

Professional body

An organisation that exists to promote or support a particular profession, such as the Law Society and the Royal Institute of British Architects.

Professional development and training

Activities that you can undertake, relevant to your job, that will increase and/or update your knowledge and skills.

Project

A comprehensive piece of work which normally involves original research and investigation either by an individual or a team. The findings and results may be presented in writing and summarised in a presentation.

Qualifications and Curriculum Authority (QCA)

The public body, responsible for maintaining and developing the national curriculum and associated assessments, tests and examinations. It also accredits and monitors qualifications in colleges and at work. You can find out more at www.qca.gov.uk

Quality assurance

In education, this is the process of continually checking that a course of study is meeting the specific requirements set down by the awarding body.

Sector Skills Councils (SSCs)

The 25 employer-led, independent organisations that are responsible for improving workforce skills in the UK by identifying skill gaps and improving learning in the workplace. Each council covers a different type of industry and develops its **National Occupational Standards**.

Semester

Many universities and colleges divide their academic year into two halves or semesters, one from September to January and one from February to July.

Seminar

A learning event between a group of students and a tutor. This may be student-led, following research into a topic which has been introduced earlier.

Specialist core units

See under **Core units**.

Study buddy

A person in your group or class who takes notes for you and keeps you informed of important developments if you are absent. You do the same in return.

Time-constrained assignment

An assessment you must complete within a fixed time limit.

Tutorial

An individual or small group meeting with your tutor at which you can discuss the work you are currently doing and other more general course issues. At an individual tutorial your progress on the course will be discussed and you can also raise any concerns or personal worries you have.

The University and Colleges Admissions Service (UCAS)

The central organisation which processes all applications for higher education courses. You pronounce this 'You-Cass'.

UCAS points

The number of points allocated by **UCAS** for the qualifications you have obtained. **HE** institutions specify how many points you need to be accepted on the courses they offer. You can find out more at www.ucas.com

Unit abstract

The summary at the start of each BTEC unit that tells you what the unit is about.

Unit content

Details about the topics covered by the unit and the knowledge and skills you need to complete it.

Unit points

The number of points you have gained when you complete a unit. These depend upon the grade you achieve (Pass, Merit or Distinction) and the size of the unit as determined by its **guided learning hours**.

Vocational qualification

A qualification which is designed to develop the specific knowledge and understanding relevant to a chosen area of work.

Work experience

Any time you spend on an employer's premises when you carry out work-based tasks as an employee but also learn about the enterprise and develop your skills and knowledge.

UNIT 2 – CONSTRUCTION AND THE ENVIRONMENT

This unit focuses on the impact that the construction and built environment sector can have on the natural environment. It provides a fundamental understanding of construction and built environment activities, techniques, processes and procedures used to protect the natural environment, and the advantages of adopting a sustainable approach to construction work. This material focuses on grading criteria P1, P2, P3 and P6 and contributes to grading criteria M1 and D2.

This unit will help you to know the important features of the natural environment that need to be protected. It will also help you to understand how the activities of the construction and built environment sector can impact on the natural environment and what can be done to reduce this impact. As part of the unit you will be given an opportunity to select sustainable construction techniques that are fit for purpose, which will further test your understanding.

Content

1) **Know the important features of the natural environment that need to be protected**

   Features: air quality; ozone quality; soil quality; natural drainage landscape; natural amenities; land use; green belts; agriculture; forestry; countryside; heritage; water (resources, quality); marine environment; wildlife; biodiversity; natural habitat.

2) **Understand how the activities of the construction and built environment sector impact on the natural environment**

   Globally: build-up of greenhouse gases causing global warming; polluting emissions to air causing acid rain; ozone depletion due to use of chlorofluorocarbons (CFCs); over-extraction (of water, fossil fuels and raw materials), increased energy consumption; electricity generation; deforestation; loss of natural habitat; reduction in biodiversity.

   Locally: air pollution by combustion products and volatile organic compounds (VOCs); polluting discharges to water by communities, industry and agriculture; contaminated land; waste disposal; existing site dereliction; comfort disturbance (traffic, smells, noise, dust and dirt); increased pressure upon existing services and infrastructure; specification of hazardous materials, eg lead and asbestos; extraction of raw materials (by drilling, mining and quarrying); electromagnetic radiation from overhead power lines; sick building syndrome.

3) **Understand how the natural environment can be protected against the activities of the construction and built environment sector**

   By legislation: relevant Acts of Parliament; UK regulations; European directives.

   By control: Health and Safety Executive (HSE); Environment Agency (EA); local authorities (eg environmental services, planning, building control departments).

   By design and specification: reduction in energy usage; minimisation of pollution; reduction in embedded energy; specification of environmentally friendly/renewable materials; reuse of existing buildings and sites.

   By management: simple environmental impact assessments (EIAs); improved management of construction sites; clear policies and objectives (eg reduction in wastage, increase in recycling, noise management, dust and dirt control); sharing of good practice; raising of awareness; communication of information.

4) Be able to select sustainable construction techniques that are fit for purpose

Fit for purpose: to meet the needs of the present without compromising the ability of future generations to meet their own needs, eg social progress that recognises the needs of everyone, effective protection of the environment, prudent use of natural resources, maintenance of high and stable levels of economic growth and employment.

Techniques: energy-based, eg reduced energy consumption, improved energy efficiency, use of renewable and alternative sources of energy; materials-based, eg specification of renewable materials, consideration of embodied energy and low-energy manufacture of materials and components; waste-based, eg producing less waste and recycling more, off-site prefabrication, modern methods of construction.

Grading criteria

P1 *identify and describe four different features of the natural environment that must be considered at the planning stage*

When planning a construction project it is important to consider the impact this can have, both during the actual construction phase and during the life of the built environment. To meet this criterion you must identify four different features of the natural environment that could be affected as a result of this process. You must consider what natural features should be identified during the planning stage of construction. You will need to describe each feature and ensure that it is not a different aspect of something you've already acknowledged.

You are not expected to show how harm may occur to the natural features you identify but your answer should describe the features that must be considered at the planning stage and, more importantly, before construction begins and it becomes possibly too late to prevent harm.

P2 *identify two different forms of global pollution and describe how each may harm the local environment*

You must provide evidence of two different forms of global pollution that can occur as a result of the construction and built environment sector. Each form of pollution you evidence must be different and your work must show how it harms the local environment. Though you are expected to show the way in which harm may occur, you are not expected to understand in detail the scientific complexity underpinning the pollution.

P3 *identify two different forms of local pollution and describe how each may harm the local environment*

You must provide evidence of two different forms of local pollution that can occur as a result of the construction and built environment sector. Each form of pollution you evidence must be different and your work must show how it harms the local area and its environment. Though you are expected to show the way in which harm may occur (especially to the local environment), you are not expected to understand in detail the scientific complexity underpinning the pollution.

M1 *assess the potential environmental impact of a proposed construction project, either real or virtual, on the local natural environment*

Here you are expected to respond to either a proposed local construction project or virtual project provided by your tutor. You will need to use the information that you have been given to assess the likely impact of the project.

In your response you will need to include evidence that refers to actual features under threat during and after the construction phases. You will also need to show a consideration of the features that must be investigated before construction begins. You are not allowed to provide a non-specific response, which means that your work must reflect the proposed project and include details from your tutor's brief.

P6 *select and describe a fit-for-purpose sustainable construction technique for each of the following issues: energy, materials and waste*

This means that you need to be able to select a sustainable technique for each of the given categories: energy, materials and waste. You will need to show how the techniques can help future generations meet their own needs. For each technique, you will need to show how it can:

- meet people's needs
- protect the environment and natural resources
- ensure stable levels of demand, investment and employment.

You will need to show how each of your chosen techniques works and describe its environmental benefits.

M3 *compare selected sustainable construction techniques in terms of relative cost and performance-in-use*

This criterion requires you to provide comparisons of sustainable construction techniques, such as the ones you chose for your P6 response. In your work you will need to consider how each technique performs in relation to its purpose, and how its outlay cost relates to its lifecycle. You do not need to provide exact figures but can use terms such as cheap, moderately expensive, and so on. You should consider how well each technique would perform in relation to the reason it would have been selected.

You will need to compare techniques selected for energy, materials and waste sustainability. You may also want to use diagrams or sketches to support your answers but the standard of presentation will not form a part of the assessment against this criterion.

D2 *justify the selection of appropriate sustainable construction techniques for a tutor-specified construction project*

For this criterion you need to select sustainable construction techniques for a given construction project. This project may be real or virtual. You will need to evidence why the techniques you've chosen are the best suited to the scenario.

In your answer you will need to show why your chosen techniques are better than the suitable alternatives. You will need to consider factors such as environmental benefits, cost, performance in use, maintenance, budget and local aesthetics as well as any specific requirements of the brief and the client. You will need to show that the benefits gained from your chosen techniques outweigh those that could be obtained from other techniques that you could have chosen.

It is important to ensure that there is clear structure in your work. Begin by introducing and identifying the problem you've been given. You then need to look at several sustainable techniques that could be used, not just your preferred ones! You must discuss and analyse the advantages and disadvantages of each technique. From there, you can then select your chosen techniques and show why they're best suited. Remember to always refer back to the project itself and use the terminology of the brief to clearly show why your techniques are best suited. Do not forget that a justification question is based on reflecting an informed opinion, and to do this you must use facts, research and available data to prove your point. Without this, you have no basis on which to form a constructive argument.

You are not allowed to provide generic responses that could be given for any project.

SECTION 1 – FEATURES OF THE NATURAL ENVIRONMENT

Aiming towards P1 – identify and describe four different features of the natural environment that must be considered at the planning stage.

ACTIVITY 1

DEFINING THE NATURAL ENVIRONMENT

The natural environment must be protected in order for all life on Earth to continue. Quite simply, should we wish to expand and develop our built environments, countries must ensure that this progress is not at the expense of our planet's valuable resources.

- The natural environment can be defined as 'the natural world which exists without human intervention'. The significance of the natural environment is that all life (even our own) depends on the Earth for its wellbeing and existence. It therefore comprises all living and non-living things that occur naturally on the planet. Examples include water, wildlife and forests.

- The built environment can be defined as 'the areas and surroundings that have been influenced and constructed with human intervention'. These components make up the physical character of any village, town or city and any supporting infrastructure. Examples include buildings, roads and parks.

Task 1

Using the internet or construction books, find five other definitions for both the built environment and the natural environment. Be sure to find definitions that give examples too.

Task 2

Using your research, create your own definitions.

Task 3

With a friend, list down the named examples you found for each environment (such as town, city, water, etc). Compare these with other class members if possible.

Task 4

Create a table like the one below. Consider whether each feature is a characteristic of the built environment or the natural environment. Tick the appropriate box. You may want to work in pairs to do this task.

Feature	Natural environment	Built environment
Wildlife		
Roads		
Forestry		
Countryside		
Natural habitat		
Tunnels		
Zoo		
Marine environment		
Ozone layer		

continued on next page

Feature	Natural environment	Built environment
Natural drainage		
Churches		
Car parks		
Bridges		
Green belts		
Playground		
Pathways		
Air quality		
Biodiversity		
Railways		
Sewers		
Water		

ACTIVITY 2

IDENTIFYING FEATURES OF THE NATURAL ENVIRONMENT

The term 'natural environment' is very likely to be new to you. The reassuring aspect, however, is that all the characteristics of natural environment are features readily available in your area or in places you may have visited in your life. This activity will now help you to identify the natural features in your area.

Task 1

List and name four examples of the natural environment in your local area. In each case, state the type of feature, its name (if known) and where it can be found (general location). It is advisable to use resources such as local plans and Ordnance Survey maps. Working with other class members could help too.

Task 2

Use Google Earth (or a similar satellite viewing application) to find four of the locations you identified. Print these aerial views and label them to show the feature type, and its name and location. For each one, try to make notes explaining what it provides (in terms of facilities, resources, life) for the surrounding area.

Task 3

Using the CAD plan shown here, identify at least three features of the natural environment.

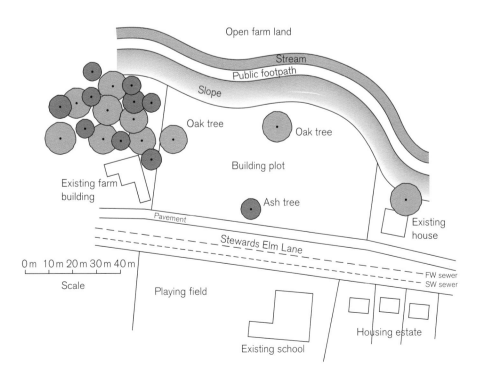

ACTIVITY 3

RESEARCHING AND DESCRIBING FEATURES OF THE NATURAL ENVIRONMENT

Natural environments are essential for the wellbeing and enjoyment of life on our planet. Their features interconnect and often rely on each other for survival. For example, wildlife will not flourish in an environment unless it has the water, food and cover that it needs to survive.

Task 1

In pairs select two features of the natural environment from this list.

- wildlife
- soil quality
- landscape
- forestry
- countryside
- natural habitat
- marine environment
- ozone levels
- natural drainage
- green belts
- air quality
- biodiversity
- water

List as many words as you can that are associated with each feature and present them as a mind map on A3 landscape paper. Consider the following aspects:

- the location of the feature
- what it provides
- what it supports (including other natural features).

An example for water would be:

- **location:** rivers, lakes, seas and aquifers
- **what it provides:** drink, washing, cleaning and bathing
- **what it supports:** health, safety, life, wildlife, forests, and marine and river life.

You may wish to use your class textbook, handouts or the internet to help with this task. Select two of the following natural features: wildlife, soil quality, landscape, forestry, countryside, natural habitat, marine environment, ozone quality, natural drainage, green belts, air quality, biodiversity, and water.

Task 2

Develop your ability to produce clear written explanations and original text. Take four natural features and create a few sentences for each, explaining where the feature can be found (use a general location), what it provides and also what it supports. You may want to use a thesaurus to help you.

Ensure that you write your sentences in such a way that they do not repeat words, and especially those used in the name of the feature. This is to help you develop original pieces of work throughout your qualification. An example has been given below; use this as a guide.

Water

This feature can be found throughout the UK and the world. It can be seen above ground in rivers, lakes and seas. It can also be found below ground level in aquifers.

This resource is essential for life on earth. It helps us to carry out our daily routines of drinking, washing, cleaning and bathing. This element ultimately helps us to stay healthy. It provides a habitat for water life and supports forests.

Task 3

Use your explanations to challenge other people. Create a small test by removing the titles from your explanations of natural features. Give these to someone else, maybe your tutor or a class member, and see if they can correctly identify the feature you have described.

For the test to work, you must ensure that you do not use the words found in the natural feature's title. If someone cannot correctly identify a natural feature from your explanation, improve your description so that the meaning is clearer.

ACTIVITY 4

WHY WE SHOULD PROTECT THE NATURAL ENVIRONMENT

Failing to protect features of the natural environment can lead to severe consequences, both for the feature itself and for human life too. These tasks will help you to identify the consequences of failing to protect the natural environment.

Task 1

Read through the lists below of features and consequences of failing to protect the environment. Using your knowledge from Activities 1, 2 and 3, match each natural feature to the most likely consequence of failing to protect that feature. You may need to carry out some further research to complete this task.

Feature	Consequence
Air quality	Flooding
Ozone layer	Imbalance and infestation
Soil quality	Extinction
Natural drainage	Over-extraction
Green belts	Reduced oxygen production
Forestry	Poor visibility
Water	Infertile land
Wildlife	Increased UV radiation
Biodiversity	Urban overspill
Natural habitat	Movement of species

Task 2

Natural features contribute to the quality of life for all living organisms. For each of the features listed in Task 1, state why the world needs the feature (or what it provides) and the consequences of destroying this feature. Present your answers in a table like the one below. You may wish to complete this table as part of a group.

Natural feature	Why the world needs this feature (what it provides)	Consequences of destroying this feature
Air quality		
Ozone layer		

Task 3

When studying for your qualification you will regularly come across new terminology. In order to prepare you for the remainder of this unit, create a glossary. Use the internet, dictionaries and construction textbooks to find definitions for the key terms listed in the table below. Copy the table so that you can keep a record of your definitions with the rest of your work. You could work in groups if you wish. It is advisable to find definitions for all the terms, but if there is little time available, try to find at least ten of them.

Key term	Definition
Biodiversity	
Air quality indicator	
pH	
Green belt	
Greenfield site	
Brownfield site	
Aquifer	
Contamination	
CO_2	
CFCs	
Arable land	
Livestock land	
Archaeological site	
Listed building	
Marine environment	
Flash flood	
Levee	
Flood plain	
Moor	
Heath	
Fenlands	
SSSI	
Residential	
Commercial	
Fertiliser	
Bund	
CDM regulations	
City	
Conurbation	
Derelict	
Environment	
Landscaping	
Pollution	
Rural	
Slum area	
Suburban	
Town	
Town planning	
TPO	
Urban	

ACTIVITY 5

HOW CONSTRUCTION AND THE BUILT ENVIRONMENT AFFECTS THE NATURAL ENVIRONMENT

The built environment can have severe consequences on the natural environment, not only during the construction process but during the life of a building too. The tasks in this activity will help you to identify these detrimental effects.

Task 1

Think about how the areas in which you live have changed over the last ten years. Consider where your home is, where you've been educated and also the villages, towns and cities that are close to you.

Task 2

Using Google Earth (or a similar application), print out aerial shots of the areas you identified in Task 1. On each printout, label the images to show where built environments have been created, developed, altered or removed over the last 10–15 years. It would useful to include the dates of any new developments. In order to improve this work, it is advisable to access your local authority's website and to consult people more familiar with these areas.

Task 3

Using the areas you've identified, make a list of the effects (both positive and negative) encountered as a result of developments. Consider both the construction and building life stages of any developments.

ACTIVITY 6

WAYS TO PROTECT THE NATURAL ENVIRONMENT

There are a range of methods available to reduce the impact of building work on the natural environment during the construction phase. Several organisations are working to reduce the environmental impact of the construction sector and similar industries. One research and information body operating in this area is CIRIA. Though created to support its members, the information provided by CIRIA is very useful to this unit.

Task 1

Go to main CIRIA website (www.ciria.org.uk) and then go to its Compliance section. (This can be found through the CIRIA website's menu.) Read the guidance section of the Compliance+ material. Make note of the various methods available to reduce the impact of construction on the natural environment.

Task 2

Using the resources available from CIRIA, find out what can be done to protect the environment during the pre-construction and construction stages of construction. Use a table like the one on the next page to record your notes. Add this to your research.

Natural feature	Pre-construction methods (see note)	Construction methods
Land		
Ecology		
Air		
Heritage		
Water		

Note: Pre-construction methods are defined as the measures used during the planning and assessment stages, before the physical construction work begins. Examples of pre-construction work include site investigations or desk-based assessment reports.

Task 3

It is important to be familiar with the protective measures and procedures demanded by legislation as well as those which are simply good practice.

Access the Environmental Impact Assessment activity found on the National Learning Network website at the address below. Type the link carefully! Take time to read the instructions and objectives, and then work through the interactive routines given.

http://go.nln.ac.uk/?req=%7BBBF652B7B-2168-4B39-85C1-6445949ED6D3%7D&prov=aclProv2264

This activity provides you with an opportunity to assess the suitability of building on either greenfield or brownfield sites, and to consider the respective advantages associated with using each type of site. It also helps you in correctly identifying natural features under threat.

SECTION 2 – FORMS OF GLOBAL POLLUTION

Aiming towards P2 – identify two different forms of global pollution and describe how each may harm the local environment.

ACTIVITY 1

UNDERSTANDING GLOBAL POLLUTION TERMINOLOGY AND ASSOCIATED SCIENTIFIC TERMS

It is important to be familiar with the terminology associated with the causes and impacts of global pollution. This activity aims to assist you to gain this understanding and improve your awareness of the basic scientific principles.

Task 1

Research the global pollution terminology given below. Write a short sentence explaining each term and present your findings as a glossary. You can work in small groups if you wish. If working on your own, select ten terms to explain.

Air conditioning	UNFCCC	Molecules	Nuclear	Habitat
Atmosphere	Sulphuric acid	Biodiversity	Detrimental	Hardwood
Climate	Sulphur dioxide	Precipitation	Distribution	Combustion
Energy	pH	UV radiation	Borehole	Infrastructure
Global warming	Methane	Prevailing	Resources	Sewage
Greenhouse gas	Carbon dioxide	Fossil fuels	Decay	Landfill
Photosynthesis	Nitric acid	Micro-organisms	Fertile	Deciduous
Raw materials	Chlorofluorocarbons	Extracted	Reservoir	Deforestation
Carbon sinks	Carbon trading	IPCC	Kyoto Protocol	

Task 2

Use the internet and environmental and scientific publications to research definitions for the term 'global pollution'. It is advisable to define each word separately too; that is, global then pollution. (Be sure to make a note of the sources and causes of global pollution you find during this process.) Present your research in the following way using A3 paper and colouring pencils.

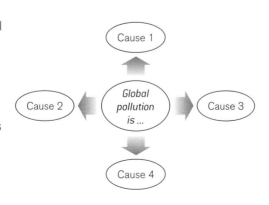

49

ACTIVITY 2

IDENTIFYING TYPES OF GLOBAL POLLUTION

In order to gain P2 it is important to identify two different sources of global pollution and link them to their impacts on the local environment. You will now work through some activities to aid your awareness of different types of global pollution.

Task 1

Briefly describe using bullet points the local impacts of each of these sources of global pollution:

- greenhouse effect
- burning fossil fuels
- deforestation
- energy consumption
- over-extracting water
- using raw materials and depleting natural resources
- habitat loss

Task 2

Global warming and acid rain are two significant global pollution issues.

- **Global warming** is caused by the excessive production of greenhouse gases as a result of industry by-products, burning fossil fuels, motor vehicle emissions and deforestation.
- **Acid rain** results from pollutants in the atmosphere, discharged through the burning of fossil fuels (to meet increasing demand for electricity) and motor vehicles emissions. These pollutants make rain more acidic, as it contains more sulphuric and nitric acid.

Read the questions given below on the greenhouse effect and acid rain. Using a variety of sources (such as your class textbook, library books, the internet, etc), try to answer every one.

Questions on the greenhouse effect

1. Describe how global warming occurs.
2. Investigate the causes given above and explain in more detail three ways in which humans are producing (ie helping to increase) greenhouse gases.
3. List four disadvantages of global warming to the environment.

Questions on acid rain

1. What makes acid rain different from normal rain?
2. Investigate the causes given above and explain in more detail three ways in which humans contribute to producing acid rain.
3. List the damage that can be caused by acid rain.

ACTIVITY 3

EXPLORING THE IMPACTS OF GLOBAL POLLUTION

Task 1

There are many websites that can provide you with the information you need in order to identify and understand the causes of global pollution and its impact on our local environment. In this task, you should use the internet to research the causes and impacts of:

- global warming and greenhouse effect
- acid rain
- ozone depletion

- over-extraction of water
- high demands on fossil fuels and raw materials
- increased energy consumption
- deforestation
- loss of natural habitat
- reduction in biodiversity.

Choose two of the topics in this list to research in depth. Then select five of the websites listed in the table below. Using these websites, establish the causes of each of your chosen topics. Then research the impact each has on your local environment (both natural and built). Present your findings as a folded A4 leaflet. Be sure to include the hyperlinks from your identified sites.

Organisation	Website
Act on CO_2	http://actonco2.direct.gov.uk/
DirectGov	www.direct.gov.uk
BBC (Bloom and Climate Change sections)	www.bbc.co.uk/bloom/
Defra	www.defra.gov.uk/
Environment and Heritage Service	www.ehsni.gov.uk/
Scottish Environment Protection Agency	www.sepa.org.uk/
NetRegs	www.netregs.gov.uk/
Friends of the Earth	www.foe.co.uk/
Greenpeace (EfficienCity section)	www.greenpeace.org.uk/files/efficiencity
Environment Agency	www.environment-agency.gov.uk/
UN Framework Convention on Climate Change	http://unfccc.int/2860.php
Government Office for London	www.gos.gov.uk/gol/
European Commission	http://ec.europa.eu/index_en.htm
CIRIA	www.ciria.org.uk/
TRADA	www.trada.co.uk/
Forestry Commission	www.forestry.gov.uk/

Task 2

Test your awareness of four significant features of global pollution. Copy Table 1 and write the causes, impacts and solutions from Table 2 in the correct columns. Use a blue pen or pencil to denote a cause, red to denote an impact and green to denote a solution. (Note: some entries in Table 2 may need to go in more than one column in Table 1.)

Table 1: Significant features of global pollution

Global warming *Causes, impact and solution*	Acid rain *Causes, impact and solution*	Ozone depletion *Causes, impact and solution*	Deforestation *Causes, impact and solution*

Table 2: Associated causes, impacts and solutions

Greenhouse gases	Decreased fish population	Desert conditions	Eye disorders
Slow growth of trees	CFCs	Increased UV radiation	Increased use of renewable energy
Killing of certain organisms	Oxides of sulphur	Rising sea levels	Nitric acid
Recycling	Forest fires	Turn your heating down	Affect biodiversity

51

Walk instead of drive	Increase in climatic temperature	Decreased filtration	Inhibited growth of green plants
Decreased micro-organisms	Skin cancer	Destruction of stone surface	Burning of fossil fuels
Increased carbon in the atmosphere	Expansion of towns	Sale of timber for buildings	Reduced absorption of moisture and water

Task 3

Pick three of the features of global pollution covered in Task 1. For each one, summarise your research by drawing a diagram like the one below.

ACTIVITY 4

IDENTIFYING THE RELATIONSHIP BETWEEN GLOBAL POLLUTION AND LOCAL EFFECTS

It is important for you to be aware of the ways in which global pollution can impact on your local area. This is because it can ultimately have a significant effect on your life in terms of work, health and vitality. These tasks will help you to identify this relationship.

Task 1

Research and find definitions for the terms 'local' and 'impact'. Find at least three explanations for each word. Use at least one book and one website.

Task 2

Use your research to create your own definition of what 'local impact' means.

Task 3

Refer back to your research for Activities 1, 2 and 3. Identify the actual local impacts that resulted from the global polluters. Include a list of these under your definition.

Task 4

Create a three-minute educational bulletin to be aired on a local radio station that highlights the impact that global pollution can have on our everyday lives. Consider how you can use language to show people the seriousness of this issue. Present this to your class.

ACTIVITY 5

LINKING THE ACTIVITIES OF THE CONSTRUCTION AND BUILT ENVIRONMENT SECTOR WITH THEIR IMPACT ON THE ENVIRONMENT

This activity will help you to identify the ways in which specific construction and built environment activities contribute to global pollution in terms of air, land and water contamination.

Task 1

The construction and built environment sector contributes to land, air and water pollution, both through its direct activities and through its use of materials produced by other industries. Carry out research to show how the construction activities and related industries listed in the table below can contribute to global pollution. These websites should assist your research:

- Environment Agency – www.environment-agency.gov.uk
- NetRegs – www.netregs.gov.uk
- NISP – www.nisp.org.uk

If you do not have access to the internet, try using your library to find relevant books and good construction magazines.

Record your results in a table like the one below. In each cell, write a short sentence explaining why contamination can be caused. Two examples have been given for you. Complete as many cells as you can.

Construction activities and related industries	Impact on the environment		
	Land	Air	Water
Power generation		As fossil fuels are burnt they release CO_2. This greenhouse gas contributes to global warming.	Chemicals released during power production can settle in water sources and change their chemical characteristics.
Manufacturing industry			
Sewage and water treatment			
Transport			
Waste management			
Construction and demolition			
Mining			
Paper industry			
Timber industry			

SECTION 3 – FORMS OF LOCAL POLLUTION

Aiming towards P3 – identify two different forms of local pollution and describe how each may harm the local environment.

ACTIVITY 1

IDENTIFYING THE CAUSES OF LOCAL POLLUTION

It is important to be aware of the pollutants present in your local area. These tasks will help you to identify, name and locate local sources of pollution.

Task 1

Use the Environment Agency website (www.environment-agency.gov.uk) to locate sources of air, land and water pollution in your area. Go to 'Your Environment' then click on the 'What's in your backyard' sections. Find examples of pollution relating to:

- fuel and power
- metal
- mineral
- chemical
- waste
- water
- radiation.

For three of these bullet points, carry out the following research.
1. Find an example of pollution in your area by using the Environment Agency website. Put your postcode in the box in the 'What's in your backyard' section.
2. Take note of the pollution shown. Find out if it is an example of general industrial pollution or an incident of significant or major pollution.
3. Gather as much data and information about this pollution as possible.
4. Print off and annotate a location map showing where this pollution originates.

ACTIVITY 2

THE IMPACT OF LOCAL POLLUTION ON PEOPLE, THE ENVIRONMENT AND THE SURROUNDING COMMUNITY

Task 1

Use the Environment Agency website to locate specific incidents of local pollution across the UK. Go to the 'Your Environment' link then the 'Your Environment magazine' section. Using the current issue, and any past issues of the magazine that are available, find an example of the construction and built environment sector causing devastation to the local environment.

When you have a found a relevant article:

- make a note of the factors that led to the pollution
- list the impacts this had on the environment and those involved
- take a brief note of any other interesting factors
- write a one-minute news report to be presented on a local television news channel — be sure to make it factual and to the point
- present your report to your class.

Note: if you do not have access to the internet, you could use copies of monthly construction and building magazines.

Task 2

Carry out research on three of these causes of local pollution:

- volatile organic compounds (kitchen cleaning products and paint)
- combustible pollutants (cigarettes, engines and fires)
- polluted rainfall entering drainage systems
- flooded sewer systems
- agriculture
- factories and industry
- waste disposal
- derelict sites
- comfort disturbance
- congested roads
- hazardous building materials
- extracting raw materials
- electromagnetic radiation from overhead power lines
- sick building syndrome.

For each chosen cause, make note of what it is and describe the impacts it can have on people and the environment.

ACTIVITY 3

REDUCING THE IMPACT OF LOCAL POLLUTION

Task 1

There many ways that local pollution can be reduced or prevented. Using your research, identify some methods that you could promote in your area to address the problem of pollution.

Present your ideas as a flyer that could be issued to your community by the local authority. Ensure that you illustrate and describe the likely impacts of pollution on people and the environment.

SECTION 4 – ASSESSING THE IMPACT OF CONSTRUCTION

Aiming towards M1 – assess the potential environmental impact of a proposed construction project, either real or virtual, on the local natural environment.

ACTIVITY 1

ASSESSING THE IMPACT OF CONSTRUCTION

Task 1

In pairs, undertake research into environmental impact assessments (EIAs) and then tackle these problems.
1. Define what an EIA is.
2. Who decides when an EIA is necessary?
3. List developments that always require an EIA.
4. List five developments that may require an EIA.
5. Locate an example of an EIA or an environmental statement.

Task 2

Using the EIAs and environmental statements you have found, discuss how they would help you to minimise the impact of a construction project.

Task 3

In groups, read the three case studies below. Select one of the case studies and then carry out these tasks.
1. Create an environmental impact assessment and environmental statement, to analyse the impact of the development, that could be presented to the developer.
2. Identify suitable methods that could be used to protect the natural environments and areas surrounding the site. Consider the impact during and after construction.
3. Present your findings and solutions to your tutor and class. Have them act as the developer of the project and ask them to mark your group's ability to assess and address the likely impact on the local environments.

Case study 1

The local council plans to transform a beleaguered and vandal-plagued open grass square and its gardens into a stunning European-style plaza. The area is currently viewed as a haven for drug-takers, and attracts crowds of congregating teens and vandals. The proposed changes will support the local authority's current master plan.

The design intention is to include a grand gateway, play areas and lawns, water fountain, café, sculptures and ornate seating. Other significant details in the plans include uplights being fixed to trees, dog-proof fencing and high-quality paving. The intention is to keep the area free from antisocial behaviour. The development is expected to cost more than a million pounds.

The plan is to use the new square for seasonal events. There will be an outdoor ice rink over the Christmas period and open-air film screenings during the summer. The area will be changed as it is currently underused. The local residents should have the benefit of a safer and cleaner environment.

Local feedback so far has been amazing and positive. Residents of the square have viewed the plans and shown great support for the design and its specifications.

By sprucing up the gardens, the council hopes to attract more residents, families and day trippers to this part of town. As it is one of the few green spaces remaining close to the town centre, it is important that its future is considered.

Funding for this ambitious development is expected to come from a mixture of lottery funding, as well as private businesses and public sector organisations.

Case study 2

Your nearest Sealife Adventure Centre could be extended. The developer's plans are to add a new children's centre, snack bar, restaurant and ecological fish farm. The centre is located at your nearest seafront and the plans involve a large-scale expansion. Not only will the centre extend on its existing site, it will also replace and use other sites surrounding it. This means a local green area will be lost.

The developer has made it very clear that the aim of the expansion is not only to make money but also to help the local marine environment too. The intention is to use the newly developed site to create a breeding programme for Dover sole that can help to restock the local waters.

Sea Associates, owners of the site, believe that the development will add a new vibrancy to the esplanade. It is also viewed as an important improvement as it will bring a new facility to the seafront. The new children's centre will be a separate two-storey building and include a drinks and snacks area and a restaurant. There will also be inside and outside play areas.

The building has been designed to a high standard by the architects. They have aimed to address any conflict with the listed fishermen's cottages on the other side of the esplanade. However, residents living in the cottages near to the site are worried about the impact of the plans on the area. They feel it will ruin their view and will wreck the last pretty area left between them and the sea. They've commented that they will just be left with a view of concrete in front of them and they'd rather have the local green kept as it is. The residents feel that the community will unfortunately lose out, despite their valid opposition.

Case study 3

A council would like to develop and widen a road close to a college and, more importantly, close to a publicly owned park. As a result, the local community and borough committee is keen to discover the likely impact that the development might have on the local environment before the road is widened.

At present the road runs along the north side of a large park that has historic and natural interest. Over 100 mature trees line the boundary between the road and the park, and these are home to many bird species and other wildlife. The road also crosses a stream which flows south into the park itself. Here it feeds two small lakes which are stocked with a range of freshwater fish. Significant information about this park has recently been unveiled, as it is believed that Natural England will soon class this site as a SSSI due to its rich variety of protected fauna and flora.

The road also travels straight through an area which was recently identified as being a site of historic interest. A Saxon burial ground was found here and it is likely that other artefacts are yet to be located.

The council would like this development to go ahead for two main reasons: it will reduce current traffic congestion and it will provide the better infrastructure vital to access a planned large shopping centre.

Major concerns about this proposal include the proposed loss of natural habitat due to the removal of 100 mature trees, the effect on biodiversity and the loss of archaeological artefacts, as well as water, land and air contamination and the direct impact on the residents who live alongside the road.

SECTION 5 – SUSTAINABLE MATERIALS

Aiming towards P6 – select and describe a fit-for-purpose sustainable construction technique for each of the following issues: energy, materials and waste.

ACTIVITY 1

UNDERSTANDING SUSTAINABILITY

In order for the construction industry to reduce its impact on the environment, sustainable building materials must be used. The term sustainable originates from the word 'sustain' and means:

- keep up
- keep from falling or giving way
- support
- bear weight of
- keep going
- give strength to
- uphold
- endure
- keep alive
- last into the future.

Task 1

In pairs, create mind maps to show the various reasons why you would want to use sustainable (environmentally friendly) techniques to build a structure. Consider why you would want to help the structure last longer; consider what the world would gain.

Discuss your ideas with other class members where possible.

Task 2

Meeting the needs of the present without compromising those of the future generations presumes that we can already meet today's requirements at an adequate or acceptable level.

Create a mind map to illustrate why we cannot always meet these needs. Consider what is wrong with today's world. It would help if you refer to a good daily newspaper to complete this task, and work with others if you can.

ACTIVITY 2

IDENTIFYING AND INVESTIGATING SUSTAINABLE TECHNIQUES

Task 1

It is important to add to the glossary you began earlier on for this unit. Research meanings of the terminology in the table below and add to the document that you created. If time restricts you, research at least ten key terms.

Refurbish	Prefabrication	Maintain	Just-in-time
Component	Reuse	Carbon neutral	Earthship
Energy efficiency	Recycle	Legislation	Reduce
Specification	Renewable	European Union	In situ
Low-energy manufacture	Alternative	Embodied energy	Waste
By-product	Renewable resource		

Task 2

Using the following websites, identify at least 10 sustainable building techniques that can be used to minimise effects on the environment. Copy and use the table below to complete this task:

Organisation/website	Sustainable building techniques
Centre for Alternative Technology www.cat.org.uk	
Sust www.sust.org	
Defra www.defra.gov.uk	
Thornhill EcoHouse web.onetel.net.uk/ ffjohndecarteret/	
Greenpeace www.greenpeace.org.uk	
National Self Build and Renovation Centre www.mykindofhome.co.uk/	
GreenSpec www.greenspec.co.uk	
Planning Portal www.planningportal.gov.uk	
ZEDfactory www.zedfactory.com	

Task 3

Sustainable construction techniques can fit into one or more of three categories.

- **Energy-based**: Techniques that reduce energy consumption, improve energy efficiency or use renewable and alternative sources of energy
- **Materials-based**: Techniques that use materials that can be renewable, require low energy levels to produce or manufacture, or that have a lower embodied energy
- **Waste-based**: Techniques that result in less waste, allow recycling, make use of off-site fabrication or use modern methods of construction to reduce by-products

Using these explanations, note which categories the construction techniques in the table below fall under. Copy the table and tick the relevant columns. You may want to refer back to the websites in Task 2 to help you.

Sustainable construction technique	Energy-based technique	Material-based technique	Waste-based technique
Grade A electrical appliances			
Green roofs			
Cast earth			
Sheep's wool recycled for insulation			
Straw bale			
Cedar timber cladding			
Redressed or crushed slate			
Hardboard windows			
Hydroelectric schemes			
Modular steel units/pods			

continued overleaf

Sustainable construction technique	Energy-based technique	Material-based technique	Waste-based technique
Heat pumps			
Low-energy light bulbs			
Combined heat and power plants			
Geothermal water heating systems			
Crushed concrete			
Recycled steel			
Engineered timber beams			
Rainwater harvesting			
Recycled concrete and brickwork			
Wind turbines			
Reused bricks			
Solar water collectors			
Photovoltaic cells			
Energy efficient boilers			
Ordering correct amount of material			
Prefabricated timber structures			
Rammed earth			
Precast concrete			

Task 4

Imagine that you are creating a booklet to illustrate three sustainable building techniques that can be incorporated within a project. Produce a brochure illustrating your chosen options. Each page should explain one sustainable building technique and must answer the following questions.

- What is the name of the technique?
- What sustainable building technique category does it fit into?
- What is this technique used for? What does it achieve for the development?
- How does it help the environment? What advantages does it have?
- How much does it cost?
- Where can this technique be used? Are there any restrictions on its use within a development?

Include any relevant sketches or diagrams.

For this task, you can use library resources, class handouts or the internet.

Task 5

Off-site construction and prefabrication can dramatically reduce waste produced during production or during a build. Investigate the following:

- a company which manufactures structures off site in controlled conditions.
- a development which has used these prefabrication units for a commercial or residential build
- the advantages for the construction process and the environment which result from this form of construction

ACTIVITY 3

IDENTIFYING SUSTAINABLE TECHNIQUES IN USE

It is important to be able to identify where sustainable techniques have been used in a development, rather than simply being informed that they're there. The tasks in this activity will help you to identify where sustainable techniques have been used.

Task 1

Read the case studies below. For each one:

- identify the sustainable building techniques used
- list the environmental advantages and state why these techniques were used.

Then select one project (case study 1 or 2) and investigate it further, using the internet or library resources. Create a detailed poster to display the project in your classroom.

Case study 1: Wales Institute for Sustainable Education

The Wales Institute for Sustainable Education (WISE) will provide an environment to educate a wide range of participants such as business leaders, policy makers, professionals and students in the principles of sustainable development. When completed, WISE will include 24 en suite study bedrooms, a 200-seat lecture theatre, workshops, seminar rooms, laboratory, restaurant and bar.

The main structure of the building will be created by a large glue-laminated (glulam) timber frame, and the lecture theatre will be built using rammed earth. All external walls will be of hemp and lime construction cast round the frame to create an insulated airtight but still breathable construction.

In order to reduce energy consumption during construction, the developers are currently gaining information about the embodied energy of the specified materials. This has meant tracing back where materials have come from, considering how they will arrive on site and monitoring how much is recyclable (as well as what is not). This will ensure that waste management is considered throughout the construction process and therefore meets current legislation requirements.

Like many sustainable buildings, the structure maximises natural daylighting and natural ventilation. It also uses:

- low embodied energy construction materials such as earth and hemp
- bio-composite, natural fibre technologies using hemp and lime
- energy efficient glazing for maximum natural daylighting and passive heat gain
- solar water heating integrated into a district heating system
- semi-transparent PV technologies used to provide both energy and shading
- biomass combined heat and power linked to the district heating system and grid
- biological, zero energy input sewage treatment systems.

Case study 2: Wessex Water building

The building's design took into account several factors, including site ecology and topography. Using the prevailing south-westerly wind, the building is able to naturally ventilate and it is oriented to maximise views of the surrounding area. This positioning also allows its offices to benefit from the winter sun in the mornings.

Being a two-storey building, the structure reduces its impact on the surrounding landscape by sitting low within the land and following the contours of the site's natural slope as best it can.

The structure's design incorporates several techniques to reduce energy usage. These include thermal mass concrete slab soffits to retain warmth in the winter, BMS-controlled vents and fan-assisted ducts for summer time night cooling, a back-up cool air system for rare hot days, and a mechanical air extract in some areas where natural ventilation is not possible.

In order to heat the office spaces a mixture of atmospheric and condensing boilers are used. This provides perimeter trench heating on all external facades and underfloor heating to the street and restaurant.

There is porous block paving on the car parking surfaces. Water percolates either to soakaways or storage tanks at the bottom of the site. These feed the irrigation system, water features and soakaways at the north end of the site. Other sustainable features include:

- energy efficient lighting and equipment
- solar water heating panels
- solar shading
- rainwater collection for lavatory flushing in the building
- integrated water management system to minimise the discharge of stormwater to the sewerage system.

Task 2

Select one of the projects listed below and use either the internet or books to find further information about the scheme. You may want to use the GreenSpec website to obtain further information on your chosen project. See the projects at: www.greenspec.co.uk/html/imagebank/imagebankcontent.html

Identify at least three sustainable building techniques used in the project you choose. Describe the advantages these had for either the environment or the project. Present your findings to the class.

- The Ecological Building Society head office
- Great Bow Yard, Somerset, a project designed by Stride Treglown
- Plas Y Mor, an award-winning day-care centre designed by PCKO Architects for Gwalia Housing Society
- Brocks Hill Environment Centre, a project by Henderson Scott Architects and the Institute of Energy and Sustainable Development
- Shorne Wood Visitor's Centre, designed by Lee Evans Partnerships, a project set in the Kent Downs at Shorne Wood Country Park
- Pines Calyx conference centre located close to the white cliffs of Dover, designed by Helionix Design
- Coopers Road, a low-rise and sustainable community in Southwark, designed by ECD Architects
- BedZED, a development of 82 homes and 18 work/live units in the London Borough of Sutton, produced for the Peabody Trust
- Kingsmead School, White Design's all-wood primary school in Northwich
- Environmental Building, BRE's offices at Garston
- Devonshire Building at the University of Newcastle, an environmental research facility designed by Dewjoc Architects
- Arup's Solihull Campus by Arup Associates
- Integer Housing, select any project by this action research network
- Downland Gridshell, part of the museum of historic buildings in Sussex, designed by Edward Cullinan Architects and Buro Happold

ACTIVITY 4

SELECTING SUSTAINABLE CONSTRUCTION TECHNIQUES

Task 1

Visit the websites of at least two providers of gas and electricity to obtain information about how domestic households can reduce energy consumption. Make a drawing of the building below and label it to show how the energy consumption in this home can be reduced.

Task 2

When a building is demolished or renovated there are materials that can be reused or recycled. Imagine that your own home (or your school or college) is going to be demolished or renovated; what items could be recycled or reused? List your answers.

Task 3

Recycling materials can provide many advantages. Research the benefits of this recycling and create a poster to display in your college or school. You may wish to show what items can be recycled from your home (or school or college).

Task 4

Sketch or take photos of your school or college. Label your images to show where the following popular sustainable techniques can be used:

- geothermal heating
- grass roof
- passive solar gain
- natural ventilation
- solar collectors
- timber glulam frame
- straw bales
- mini wind turbine
- rainwater harvesting
- rammed earth
- precast concrete
- sheep's wool insulation

Briefly describe the environmental benefits of each technique.

Task 5

Find diagrams of these energy-conserving sustainable building techniques.

- solar collector (flat plate collector)
- mini wind turbine
- small-scale hydroelectric power

Make a sketch of each diagram and label it appropriately. Be sure that your sketch shows all components and indicates how energy is produced.

Task 6

Six key attributes are vital to the success of any sustainable project.
1. Identifying appropriate targets and using correct management techniques
2. Supporting communities
3. Enhancing biodiversity
4. Creating a healthy environment
5. Using resources effectively
6. Minimising pollution

Create a sustainable design solution to suit the following design brief.

Design brief
- Produce a sustainable development on the edge of a small town
- Construct seven large detached two-storey houses (with integral double garages)
- Protect existing trees on the site
- Address the poor drainage on the site

- Promote, protect and create (where possible) the natural environment
- Create a communal park
- Create pedestrianised streets
- Promote natural habitats

This is a location plan of the site for the development

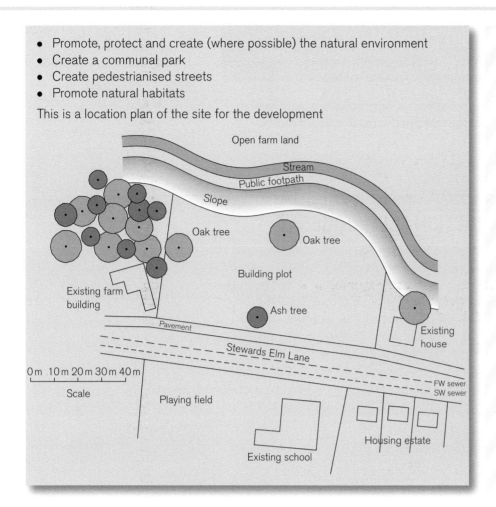

Present your ideas using suitable drawings. Provide an overview of the sustainable materials you've selected, and identify at least five sustainable building techniques that can be used on this site. You must have at least one energy-based technique, one material-based technique and one waste-based technique. Discuss why you have chosen each technique.

Task 7

Find out about the Code for Sustainable Homes. As part of your research identify:
- key dates
- the purpose of the rating
- establishments and bodies associated with the code's enforcement
- what the code has replaced
- how a home can be rated
- how this will affect home owners and developers.

Task 8

Some sustainable building techniques could be considered more suitable for small private residential buildings than for larger commercial projects. This is mainly due to the different cost and complexity of particular methods.
In small groups, identify three techniques that you would normally associate with a small sustainable residential building and three you would expect to find on a larger more complicated build. Present your reasons to the class.

SECTION 6 – JUSTIFYING SUSTAINABLE TECHNIQUES

Aiming towards D2 – justify the selection of appropriate sustainable construction techniques for a tutor-specified construction project.

ACTIVITY 1

JUSTIFYING THE SELECTION OF APPROPRIATE SUSTAINABLE CONSTRUCTION TECHNIQUES

In groups, read each scenario brief and then choose one of the projects. Your task is to create a sustainable design solution for your chosen project. Produce a set of drawings to present your ideas. (These could be either 2D or 3D, or both.) Present your ideas on A2 paper and, for each sustainable technique chosen, provide a brief explanation (perhaps as a set of bullet points) of why this technique is more effective than possible alternatives. Provide a brief list of the alternative techniques that you considered but have not chosen. Present your design solution to your class.

Try to show consideration to:

- local communities
- the construction process and the building's life
- the requirements of the brief and any targets set
- creating and maintaining a healthy environment
- using resources effectively
- minimising pollution
- enhancing biodiversity.

Scenario 1

Your college or school would like to design and build a new facility in which to deliver its construction courses. The site is intended to be used for both theory and practical lessons (craft, technical and professional).

The facility must include:

- five areas suitable for teaching construction theory
- a workshop to allow craft activities to be taught, including bricklaying, carpentry and joinery, plumbing, painting and decorating
- a staff room for teaching staff
- suitable storage areas for technical and professional equipment
- a canteen, toilets and changing rooms
- a main reception and visitor waiting area.
- car parking for staff, learners and visitors.

Scenario 2

Your local authority would like you to create a new visitor centre in your area. It is hoped that with this building will encourage more tourists to visit the area and help to boost the local economy.

The council has requested that the visitor centre meets these requirements:

- uses natural materials for its exterior and structure
- requires low levels of heating, electricity and water usage
- uses at least three local materials, products or techniques
- includes accommodation for a small café, information display area, front reception, toilets and main office for three members of staff
- matches local aesthetics.

ANSWERS

SECTION 1

ACTIVITY 1

Tasks 1–3

There are no specific answers to these tasks.

Task 4

Feature	Natural Environment	Built Environment
Wildlife	✓	
Roads		✓
Forest	✓	
Countryside	✓	
Natural habitat	✓	
Tunnels		✓
Zoo		✓
Marine environment	✓	
Ozone layer	✓	
Natural drainage	✓	
Churches		✓
Car parks		✓
Bridges		✓
Green belts	✓	
Playground		✓
Pathways		✓
Air quality	✓	
Biodiversity	✓	
Railways		✓
Sewers		✓
Water	✓	

ACTIVITY 2

Task 1

There are no specific answers to this task. You are expected to find local examples of the natural environment. These should include local examples of landscape or countryside that can be distinguished in different ways, such as woodlands, rivers, streams and lakes, fenlands, moors, heaths, meadows, mountains, seas, estuaries, cliffs and beaches.

Task 2

There are no specific answers to this task. You must find satellite images to illustrate your local environment. You should note what these provide for the area, such as habitat, resource, biodiversity, life source, etc.

Task 3

Features to note include:

- woodland
- stream
- natural drainage
- open farmland.

You should also note that wildlife species will occupy the habitats shown on the plan.

ACTIVITY 3

Task 1

Here are answers for some of the features in the list.

Wildlife

- Location: Parks, forests, woodlands, fields, gardens, heaths, meadows, moorlands, etc.
- What is provides: Biodiversity, balance of species, soil nutrients, food source.
- What it supports: Food chains, predators, biodiversity, soil fertility, migration of seeds, plant reproduction, recreation activities, etc.

Natural habitat

- Location: Trees, soil, sand, plant life, parks, fields, gardens, heaths, meadows, moorlands, etc.
- What is provides: Protection, safety, nurture, food source, building resource, fuel.
- What it supports: Survival, development and existence of wildlife, reproduction and biodiversity.

Natural drainage

- Location: All soil and exposed land surface found in river levees, woodland areas, greenfield or green belt areas.
- What is provides: Natural control of water levels, channels our water resources, drains the land, avoids flooding and washes nutrients from the land.
- What it supports: Control of nutrient and disease, movement of species, plant growth, human management of land and water.

Task 2

Your answers should expand on key words and phrases used in Task 1 responses.

Task 3

There are no specific answers to this task. You must use correct terminology without stating the natural feature's actual name.

ACTIVITY 4

Task 1

Feature	Consequence
Air quality	Poor visibility and spread of disease
Ozone layer	Increased UV radiation
Soil quality	Infertile land
Natural drainage	Flooding
Green belts	Urban overspill
Forestry	Reduced oxygen production
Water	Over extraction
Wildlife	Extinction
Biodiversity	Imbalance and infestation
Natural habitat	Movement of species

Task 2

There are no specific answers to this task, but the table gives some examples of what could be produced.

Natural feature	Why the world needs this feature (what it provides)	Consequence of destroying this feature
Air quality	To breathe, maintain health, and to increase/sustain vision and sight.	Lack of good quality air will lead to spread of disease, breathing difficulties and even death.
Ozone layer	Filters UV rays and enables life on Earth.	Higher risk of skin cancer, damage to sight and possible death.
Soil quality	Provides fertile land for agriculture and plant life, sustains life in general, provides area for development and areas for communities to interact.	Less forestry, agriculture and general plant life, resulting in illness and death, wastelands and derelict sites.
Natural drainage	To avoid flooding, naturally control water levels, manage our water resources, drain the land, and wash nutrients from the land.	Flooding, damage to homes, spread of disease, uncontrollable movement of water and possible reduction of sources of drinking water.
Green belts	To prevent urban sprawl and overcrowding, provide a natural amenity, control development and protect natural habitats.	More pollution, derelict sites, loss of aesthetic and attractive areas, fewer clean and fresh spaces, fewer areas for communities to enjoy away from cities and towns, and uncontrollable built environments.
Water	Drinking, washing, cleaning and bathing. More importantly, it provides an area for biodiversity, a balance of species and supports food production and industrial processes.	Poor health of species (human and water-based), lack of biodiversity and lack of an essential resource.
Wildlife	Biodiversity, balance of species, soil nutrients and food source.	Negative impact on food chains, predators, biodiversity, soil fertility, migration of seeds and plant reproduction.
Natural habitat	Protection, safety, nurture, food source, building resource and fuel.	Significant impact on biodiversity owing to restricting movement of species, lack of reproduction, change in food source, changes in shelter and warmth which can eventually lead to extinction.

69

Task 3

There are no specific answers to this task. You must gather definitions that provide good detail and which you can understand.

ACTIVITY 5

Tasks 1 and 2

There are no specific answers to these tasks.

Task 3

Your answers are likely to feature some of these issues.

Positive impacts	Negative impacts
Increased economic activity	Fewer natural resources
Better amenities and facilities	Derelict land and buildings
Better and more suitable housing	Increased costs
Cleaner areas and better health	Vandalism and crime
Investment	More waste
Durable structures	Air, land and water pollution
Employment	Unemployment

ACTIVITY 6

Task 1

There are no specific answers to this task; you must read the website given.

Task 2

Here are some of the methods you may have identified.

Natural feature	Pre-construction methods	Construction methods
Land	<ul><li>Assess land for contaminants using a site soil survey</li><li>Check for archaeological sites, such as burial grounds</li></ul>	<ul><li>Deal with land contaminants appropriately</li><li>Do not stockpile contaminated soil unless it cannot be avoided</li><li>Cover over stockpile material</li><li>Control surface drainage from stockpiled areas.</li><li>Be careful when handling, storing and using oils and chemicals</li></ul>

continued on next page

Natural feature	re-construction methods	onstruction methods
Ecology	• Determine and identify designated ecological sites or protected species • Identify, fence off, and make staff aware of, any sensitive areas before works commence • Plan transport routes • Schedule work in relation to wildlife breeding or nesting periods • Establish close relationships with nature conservation bodies and local environmental groups • Plan and design to replace habitats and wildlife destroyed or removed • Check for tree preservation orders	• Keep vehicles and plant away from trees • Put up temporary fencing to mark out protected areas • Wrap damp sacking around exposed roots until backfilling • Do not store spoil or building materials within protected areas • Place protective bunds around ponds • Monitor water levels during works • Direct run-off away from sensitive areas

Air	• Plan
	  – storage sites for fuel and materials
	  – maintenance areas
	  – delivery and haulage routes
	  – construction times and seasons
	  – waste storage solutions and methods of disposal
	  – location of welfare facilities
		• Provide a length of paved road before the exit from the site
		• Sweep paved access roads using a vacuum sweeper
		• Limit vehicle speeds
		• Damp down dust
		• Use enclosed chutes
		• Locate crushing plant away from sensitive sites
		• Clean the wheels of vehicles leaving the site
		• Ensure that vehicle exhausts do not discharge directly at the ground
		• Locate stockpiles of fine-grained material out of the wind
		• Minimise the storage time of materials on site
		• Ensure that all dust-generating materials transported to and from the site are covered by tarpaulin
		• Minimise cutting and grinding on site

Methods for protecting the other natural features covered in this task – heritage and water – can be found on CIRIA's website.

Task 3

Answers and recommended solutions can be found within the activity on the National Learning Network's website.

71

SECTION 2

ACTIVITY 1

Task 1

Air conditioning: A system for controlling the temperature, condition and humidity of the air in a building.

Atmosphere: Gases surrounding and helping to protect the Earth.

Biodiversity: Range of biological species present within the environment.

Borehole: A hole that is drilled into the earth, as in exploratory well drilling or in building construction.

Carbon dioxide: A colourless, odourless, incombustible gas (CO_2) formed during respiration, combustion and organic decomposition.

Carbon sinks: A sink is a component of the carbon cycle that stores more carbon than it emits to the atmosphere. A sink can be likened to a water well. Forests and soils can be carbon sinks.

Carbon trading: A system where companies are given a limit as to how much pollution they generate. If they need to produce more, they 'buy' the surplus from a company that isn't using all their allowance.

Chlorofluorocarbons: Chemical substances used in refrigerators, air conditioners and solvents, which drift to the upper stratosphere and dissociate. Chlorine released by chlorofluorocarbons (CFCs) reacts with ozone, eroding the ozone layer.

Climate: The meteorological conditions, including temperature, precipitation and wind, that characteristically prevail in a particular region.

Combustion: The process of converting fuel into heat; requires oxygen.

Decay: Decline or decomposition or matter.

Deciduous: Shedding or losing foliage at the end of the growing season.

Deforestation: The cutting down or destruction of forests.

Detrimental: Causing damage, harm or negative effect.

Distribution: The act of spreading or apportioning.

Energy: Activity, strength, efficiency, forcefulness, intensity.

Extracted: To draw or pull out, often with great force or effort.

Fertile: Capable of producing offspring or bearing crops.

Fossil fuels: A hydrocarbon deposit, such as petroleum, coal or natural gas, derived from living matter of a previous geologic time and used for fuel.

Global warming: An increase of the average temperature of the Earth's atmosphere by a few degrees resulting in an increase in the melting of ice which contributes to sea-level rise.

Greenhouse gas: Any gas that absorbs infrared radiation in the atmosphere.

Habitat: The area or environment where an organism or ecological community normally lives or occurs.

Hardwood: The wood of a deciduous tree.

Infrastructure: Basic structure of an organisation or economic system.

IPCC: Stands for Intergovernmental Panel on Climate Change, an organisation that analyses research into climate change.

Kyoto Protocol: An agreement that commits the countries that sign up to it to reduce their greenhouse gas emissions to specific levels.

Landfill: A method of solid waste disposal in which refuse is buried between layers of soil so as to fill in or reclaim low-lying ground.

Methane: A colourless odourless flammable gas, the main constituent of natural gas.

Micro-organisms: Any organism of microscopic size.

Molecules: The smallest particle of a substance that retains the chemical and physical properties of the substance and is composed of two or more atoms; a group of like or different atoms held together by chemical forces.

Nitric acid: A clear, colourless to yellow liquid that is very corrosive and can dissolve most metals. It is used to make fertilizers, explosives, dyes and rocket fuels.

Nuclear: A spontaneous nuclear transformation (radioactivity) characterised by the emission of energy and matter from the nucleus of the atom.

pH: A measure of the acidity or alkalinity of a solution.

Photosynthesis: Process of using energy in sunlight to convert water and carbon dioxide into carbohydrates, with oxygen as a by-product.

Precipitation: Any form of water, such as rain, snow, sleet or hail, that falls to the Earth's surface.

Prevailing: Most frequent or common; predominant.

Raw materials: An unprocessed natural product used in manufacture.

Reservoir: Enclosed area for the storage of water.

Resources: Funds, means, supplies, wealth.

Sewage: Refuse and waste matter.

Sulphur dioxide: A colourless toxic gas (SO_2) that occurs in the gases from volcanoes; used in many manufacturing processes and present in industrial emissions; causes acid rain.

Sulphuric acid: A highly corrosive acid made from sulphur dioxide.

UNFCCC: The United Nations Framework Convention on Climate Change, a treaty signed at the Earth Summit in Rio de Janeiro in 1992 that aims to stabilise the amount of greenhouse gas in the atmosphere.

UV radiation: Ultraviolet radiation. Invisible rays that are part of the energy that comes from the sun.

Task 2

Global: Of, relating to, or involving the entire Earth; worldwide.

Pollution: The act or process of polluting or the state of being polluted, especially the contamination of soil, water or the atmosphere by the discharge of harmful substances.

Examples of global pollution include acid rain, deforestation, ozone depletion, over extraction of water, burning fossil fuels and loss of natural habitat.

ACTIVITY 2

Task 1

There are no specific answers to this task.

Task 2

Questions on the greenhouse effect

1. The by-products of industry, the burning of fossil fuels and the effect of deforestation all increase the volume of carbon dioxide and other greenhouse gases in the atmosphere. These gases trap heat and reflect it back down to the earth. This increases the overall temperature and causes global warming.
2. Human beings contribute to global warming through burning fossil fuels, using motor vehicles and cutting down rainforests. You should expand on these answers by providing more detail.
3. Global warming is harmful to ecosystems, plants and animals; it increases sea levels; it increases the likelihood of storm events; and it affects people's health.

Questions on acid rain

1. This is more acidic than normal rain. It contains more sulphuric and nitric acids, which come from pollutants in the atmosphere.
2. Human beings contribute to acid rain by the burning of fossil fuels, by using more electricity (as pollutants are released into the atmosphere in much electricity generation) and by using motor vehicles. You should expand on these answers by providing more detail.
3. Acid rain causes damage to trees; it pollutes freshwaters and it negatively impacts on fish and marine life populations.

ACTIVITY 3

Task 1

There are no specific answers to this task.

Task 2

Global warming *Causes, impacts and solutions*	Acid rain *Causes, impacts and solutions*	Ozone depletion *Causes, impacts and solutions*	Deforestation *Causes, impacts and solutions*
Causes • Greenhouse gases • Burning of fossil fuels	**Causes** • Oxides of sulphur • Nitric acid • Burning of fossil fuels	**Causes** • CFCs	**Causes** • Forest fires • Sale of timber for buildings • Expansion of towns
Impacts • Killing of certain organisms • Increased sea levels • Increase in climatic temperature	**Impacts** • Slow growth of trees • Decrease in fish populations • Destroys stone surfaces	**Impacts** • Eye disorders • Killing of certain organisms • Skin cancer • Affects biodiversity • Increased UV radiation • Decreased filtration • Inhibited growth of green plants	**Impacts** • Decreased micro-organisms • Reduced absorption of water • Increased carbon in the atmosphere • Desert conditions
Solutions • Recycling • Turn your heating down • Increase use of renewable energy	**Solutions** • Increase use of renewable energy • Recycling • Walk instead of drive	**Solutions** • Increase use of renewable energy • Recycling	**Solutions** • Recycling • Increase use of renewable energy

Task 3

Here are some possible answers.

Global warming	
Causes	**Impact**
Burning fossil fuels	Increased temperature
Using motor vehicles	Increased sea level
Deforestation	Harm to ecosystems

Ozone depletion	
Causes	**Impact**
CFCs	Eye disorders
	Skin cancer
	Increased UV radiation

Acid rain	
Causes	**Impact**
Burning fossil fuels	Tree damage
Use of motor vehicles	Decline in water life
Increased demand for electricity	Damage to stone surfaces

74

Deforestation	
Causes	**Impact**
Forest fires	Reduced water absorption
Increased demand for timber	Increased carbon levels
Expansion of towns and cities	Decrease in oxygen
Changing use of land	Extinction of species

ACTIVITY 4

Task 1

Local: Confined to one part, relating to, or characteristic of a particular place; relating to or applicable to or concerned with the administration of a city or town or district rather than a larger area.

Impact: Profound effect; a measure of the effect that an incident, problem or change is having or might have; the effect or impression of one thing on another.

Task 2

There are no specific answers to this task.

Task 3

Here are some possible answers.

Global pollution	Local effect
Global warming	Increased chance of flooding
Acid rain	Breakdown of stone on buildings. More acidic soil making it less fertile
Ozone layer	Skin cancer Eye disorders
Deforestation	Less oxygen produced Loss of local wildlife Flooding due to less water absorption

Task 4

There are no specific answers to this task.

ACTIVITY 5

There are no specific answers to this activity.

SECTION 3

ACTIVITY 1

Task 1

There are no specific answers to this task.

ACTIVITY 2

There are no specific answers to this activity.

ACTIVITY 3

Task 1

These are some methods of reducing the impact of pollution which you could feature in your leaflet:

- walk instead of driving
- turn heating down
- not leaving appliances on stand-by
- buy energy-efficient appliances
- use public transport
- recycle and reuse
- regular maintenance of vehicles, machinery and buildings
- plan construction sites effectively
- identify and protect sites of ecological interest
- control on-site waste
- refuel vehicles in designated areas
- do not fly tip
- report polluting offenders

SECTION 4

ACTIVITY 1

Task 1

1. An environmental impact assessment (EIA) is a process in which information about the positive and negative environmental effects of a development are considered. The intended outcome is to create an environmental statement that summarises the result of the analysis and shows the severity and likelihood of environmental impact of the proposed project.
2. Although an EU directive stipulates which specific projects require an EIA, local authorities normally consider whether an environmental statement is needed and what its scope should be. There are two schedules governing this process: Schedule 1 and Schedule 2.
3. Schedule 1 developments always require an EIA. These include oil refineries; thermal and nuclear power stations, and nuclear fuel reprocessing plants; iron and steel smelting plants; asbestos extraction installations; industrial chemical installations; major railway lines, motorways and airports; inland waterways and ports; waste disposal and water transfer or treatment plants; petroleum and natural gas extraction facilities; dams; pipelines; intensive poultry and pig installations; industrial timber and paper plants; quarries and opencast mining; and petroleum or chemical storage.
4. Schedule 2 developments may require an EIA. These include developments in relation to agriculture and aquaculture; extractive industry; energy industry; production and processing of materials; mineral industry; food industry; textile, leather, wood and paper industries; rubber industry; infrastructure projects; tourism and leisure.
5. There are no specific answers to this task.

Task 2

There are no specific answers to this task.

Task 3

You must note the key natural features and factors described in each case study and use these to suggest suitable pre-construction, construction and post-construction measures.

Responses to case study 1 should note:
- it is one of the few remaining green spaces in the local area
- the impact on the natural drainage
- the impact on the local wildlife, including habitats and biodiversity
- the negative impact of uplighters being fixed to trees and dog-proof fencing
- possible contamination due to the ice rink and water feature
- the increased waste from users
- the noise, visual and air pollution impact on local residents, both during development (due to construction work) and after construction (ice rink, open air cinema, etc).

Responses to case study 2 should note:
- possible water contamination due to the proximity of the sea in relation to the site could impact on marine life
- the loss of the local green – this could impact on biodiversity, local habitats and wildlife species
- air, sound and visual pollution due to demolition of local structures
- air, sound and visual pollution due to life of proposed new structures
- positive impact on Dover sole population

- positive economic impact
- after construction, the poor visual aesthetics, sound pollution and general waste could have negative impact on local community.

Responses to case study 3 should note:

- the site is owned by the local people – it is theirs to control
- the mature trees will be removed, with a likely negative impact on habitats and biodiversity
- the possible loss or damage of archaeological artefacts
- there is likely to be water, land and air contamination, with a consequent impact on habitats, species and biodiversity
- the residents who live alongside the road will face the inconvenience of noise, air and light pollution both during and after construction works
- significant loss of fauna and flora and threat to SSSI status
- improved traffic flow and economic investment.

SECTION 5

ACTIVITY 1

Task 1

Reasons for using sustainable (environmentally friendly) techniques to build a structure include:

- reduce speed of consumption
- to reduce/minimise costs
- avoid/reduce environmental impact
- avoid/reduce maintenance and refurbishment
- sustain jobs and/or create employment
- reduce energy consumption
- make finite resources last longer
- avoid pollution to air, land and water
- achieve durable structures
- maintain and save natural resources
- avoid sending unnecessary waste to landfill.

Task 2

There are several reasons why the world struggles to meet the needs of the present. These can be broken down into social, political, economic and environmental issues.

Social issues include:

- fashion and pace of change
- broken communities
- poverty
- inadequate technology.

Political issues include:

- war and famine
- government policies.

Economic issues include:

- investment
- unemployment
- demand
- cost of production and development
- strength of currency
- scarcity of resources
- poor or inadequate infrastructure
- government policies.

Environmental issues include:

- pollution
- extinction
- removal of habits
- lack of resources
- over-development
- government policies.

ACTIVITY 2

Task 1

You must correctly define the terminology given.

Task 2

Energy-based sustainable building techniques include wind farming, providing solar panels and installing combined heat and power plants.

Material-based sustainable building techniques including using natural insulation such as sheep's wool or Warmcel, local materials such as slate and English larch, and natural products such as rammed earth, earth blocks and green roofs.

Waste-based sustainable building techniques include greater use of local materials as well as more emphasis on recycling, reusing and prefabrication. They also cover rainwater harvesting, the use of composters and recycling hardcore from demolished buildings.

Task 3

Sustainable construction technique	Energy-based technique	Material-based technique	Waste-based technique
Grade A electrical appliances	✓		
Green roofs		✓	
Cast earth		✓	
Sheep's wool recycled for insulation	✓	✓	
Straw bale		✓	
Cedar timber cladding		✓	
Redressed or crushed slate		✓	✓
Hardboard windows		✓	
Hydroelectric schemes	✓		
Modular steel units/pods			✓
Heat pumps	✓		
Low-energy light bulbs	✓		
Combined heat and power plants	✓		
Geothermal water heating systems	✓		
Crushed concrete		✓	✓
Recycled steel		✓	✓
Engineered timber beams		✓	
Rainwater harvesting		✓	✓
Careful packaging		✓	✓
Recycled concrete and brickwork		✓	
Wind turbines	✓		
Reused bricks		✓	✓
Solar water collectors	✓		
Photovoltaic cells	✓		
Energy efficient boilers	✓		
Ordering correct amount of material			✓
Prefabricated timber structures		✓	✓
Rammed earth		✓	
Precast concrete		✓	✓

Task 4

There are no specific answers to this task.

80

Task 5

Companies that manufacture structures off site include Yorkon and the Elliott Group. Other examples can be found on the Modular and Portable Building Association website (mpba.biz).

Many projects use prefabricated units include residential, commercial, retail, leisure, education and defence developments.

The advantages of prefabrication for the environment include:
- it uses less energy than is required for on-site methods
- it can reduce the embodied energy in a building and demand on natural resources
- it can create less air pollution, noise and debris
- road traffic to and from the site is reduced and this results in less carbon emission
- it can promote more sustainable methods, such as using local sources of timber
- it leads to the efficient use of materials, labour and energy
- its enables waste to be recycled, avoiding possible air, land and water pollution.

The advantages of prefabrication for the construction process include:
- units are built in quality-controlled factory conditions
- it helps improve efficiency on site
- enables repetitive processes to become more effective and better controlled
- significantly easier to create several modules off site than it is to repeat the process on site
- it reduces the need for labour-intensive and costly alternatives
- it can address any gap in the availability of skilled workers
- it is a cost-effective solution for contractors
- it addresses some health and safety concerns on site
- it can save significant time and cost during the installation process.

ACTIVITY 3

Task 1

Case study 1

Sustainable building techniques include rammed earth, glulam timber frame, hemp and lime render, passive solar gain, solar water heating, photovoltaic cells, biomass combined heat and power system.

These provide several environmental advantages. With good insulation, breathable construction, renewable building resources, and effective heating and cooling systems, the WISE facility makes effective use of natural resources and has taken significant steps to reduce its carbon footprint and its energy consumption.

Case study 2

Sustainable building techniques include south-west facing orientation, passive solar gain, aesthetic placement within the site's contours, high thermal mass, BMS-controlled vents, natural ventilation, condensing boilers, underfloor heating, solar water heating, solar shading, rainwater collection, natural irrigation and on-site water management.

This will result in reduced energy consumption, minimal demand on natural resources (water and fossil fuels), good water management (minimising any risk of flooding), a reduced carbon footprint, minimal visual impact, consideration to surrounding environment, and a regulated internal environment promoting constant ambience and good health.

Task 2

There are no specific answers to this task.

ACTIVITY 4

Task 1

Steps that can be taken to reduce energy consumption in this house include:

- adding cavity insulation
- keeping curtains closed
- improving (increase or replace) loft insulation
- installing double-glazed windows
- using energy-saving light bulbs
- installing grade A kitchen appliances
- switching off electrical goods rather than leaving them on stand-by.

Task 2

Materials and features of a building that can be reused or recycled include the guttering, chimney pots, structural components (RSJs, corrugated iron, bricks, stone), fittings (hinges, locks, handles), roof materials (shingles, slates, tiles), fireplace items, copper, brass, lead, timber (doors, window frames, structural timbers). Did you think of any others?

Task 3

The benefits of this recycling include:

- ensures conservation of materials and allows preservation of buildings
- provides a supply of authentic period materials that are not otherwise available, such as fireplaces, doors, marble and special bricks
- reduces the amount of waste
- enhances visual appeal of new developments to match existing structures or local environment
- saves energy during the whole production cycle
- building controls may require reuse of materials for environmental or aesthetic reasons.

Task 4

There are no specific answers to this task. You must suggest the use of techniques in locations that are 'fit for purpose'.

Task 5

There are no specific answers to this task.

Task 6

Your proposed design solutions could use many different sustainable building techniques:

- Energy-based techniques include mini wind turbines, roof-mounted solar collectors and photovoltaic cells, south-facing homes, grade A electrical appliances, geothermal water heating systems, energy-efficient boilers, naturally ventilated homes with high levels of thermal performance, wind and solar powered street lighting and signage, onsite car charging points, and bus stop on route to local town (if possible).

- Material-based techniques include natural insulation, local materials such as slate and English larch, natural products such as earth blocks or green roofs, engineered timber structures, recycled concrete and brickwork, and permeable paving to help natural drainage.

- Waste-based techniques include use of local materials (as above), recycling points, reusing rainwater for flushing toilets etc through rainwater harvesting, prefabrication for proposed homes, garage units or the communal park (depending on integrity), composters (for kitchen waste or used as toilet system), hardcore from demolished buildings, careful ordering of materials, precast concrete for building elements (wall, floor and stairwells).

It is important that you choose techniques that complement each other. For example, you cannot specify precast concrete components if you have already specified a prefabricated timber design which will already have included these building elements in the manufacturing process. You should explain and indicate why techniques are suitable. Reference back to the brief should be made as much as possible, and to any site plan. Drawings could be produced to show consideration to the existing local features – the woodland, stream, embankment, footpath and open farmland.

Tasks 7–8

There are no specific answers to these tasks.

SECTION 6

ACTIVITY 1

Task 1

There are no specific answers to this task. You must try to use and justify a good range of techniques.

Suitable examples for scenario 1 could include energy-efficient lighting and equipment, solar water heating panels, solar shading, rainwater collection for lavatory flushing in the building, harvested water irrigation system, use of local materials, natural insulation, timber frame and natural rendering products.

Suitable examples for scenario 2 could include a similar selection of techniques. However, energy-based techniques should be small or discrete. Materials should be predominantly natural for example timber, earth or stone. The building's design and materials should suit the local aesthetics of the area.

UNIT 3 – MATHEMATICS IN CONSTRUCTION AND THE BUILT ENVIRONMENT

This unit focuses on the mathematical calculations that can arise during a construction project. They can take on many forms, including calculating the dimensions of a structure, determining how much of a particular material must be ordered, calculating costs, and setting out dimensions and angles. This section focuses on grading criteria P1, P2, P3, P4, P5, P6, P7; M1, M2, M3 and aspects of D1 and D2.

This unit will help you gain basic underpinning mathematical techniques and enable you to select and correctly apply them to construction problems involving perimeters, areas and volumes. You will also look at how practical construction problems can be solved by using a variety of geometric and trigonometric techniques or by using graphical and statistical techniques.

Content

1) **Know the basic underpinning mathematical techniques and methods used to manipulate and/or solve formulae, equations and algebraic expressions**

   Mathematical techniques and methods: mathematical operators; factorisation; expansion; transposition; substitution and elimination; rounding; decimal places; significant figures; approximation; truncation errors and accuracy; calculator functions and use.

   Formulae, equations and algebraic expressions: linear, simultaneous and quadratic equations; arithmetic progressions; binomial theorem.

2) **Be able to select and correctly apply mathematical techniques to solve practical construction problems involving perimeters, areas and volumes**

   Perimeters, areas and volumes: calculations both for simple and compound shapes, eg rectangles, trapeziums, triangles, prisms, circles, spheres, pyramids, cones and both regular and irregular surface areas and volumes.

   Mathematical techniques: simple mensuration formulae and numerical integration methods (mid-ordinate rule; trapezoidal rule; Simpson's rule).

3) **Be able to select and correctly apply a variety of geometric and trigonometric techniques to solve practical construction problems**

   Geometric techniques: properties of points, lines, angles, curves and planes; Pythagoras' rule; radians; arc lengths and areas of sectors.

   Trigonometric techniques: sine, cosine, tangent ratios; sine rule; cosine rule; triangle area rules.

4) **Be able to select and correctly apply a variety of graphical and statistical techniques to solve practical construction problems**

   Graphical techniques: Cartesian and polar coordinates; intersections of graph lines with axes; gradients of straight lines and curves; equations of graphs; areas under graphs; solution of simultaneous and quadratic equations.

   Statistical techniques: processing large groups of data to achieve mean, median, mode and standard deviation; cumulative frequency, quartiles, quartile range; methods of visual presentation.

Grading criteria

P1 *use the main functions of a scientific calculator to perform calculations and apply manual checks to results*

This means that you must be able to use not just the 1, 2, 3 and 4 keys etc, but also the sin, cos, tan, x^2, $\sqrt{x}$, π keys etc. You should be able to do a rough calculation to check that your answer is approximately correct, to check, for example, that you do not have a decimal point in the wrong place. Don't forget that all solutions should be set out clearly and include the proper units.

P2 *use standard mathematical manipulation techniques to simplify expressions and solve a variety of linear formulae*

The term 'linear formula' is used to describe any simple equation that has one unknown value that you want to solve. There is no need to memorise equations. However, there are two important skills that you need to have.

First, you need to be able to simplify equations in order to make the calculations easier. Second, you need to be able to rearrange equations. For example, there is an equation to convert temperature from Fahrenheit to Celsius. You should be able to rearrange this equation to produce an equation to convert Celsius to Fahrenheit.

P3 *use graphical methods to solve linear and quadratic equations*

It is possible to solve equations by drawing a graph. Therefore you need to know how to draw graphs, using suitable axes with clear labelling. The advantage of drawing graphs is that it cuts out the calculations. For example, you could have a graph to convert feet to metres. Incidentally, a linear equation will produce a straight line and a quadratic equation will produce a symmetrical curve.

P4 *produce clear and accurate answers to a variety of problems associated with simple perimeters, areas and volumes*

The most common calculations you will undertake in construction involve areas and volumes. So when an architect has designed a building you should be able to calculate the quantities of, say, concrete and bricks required to construct the building. In addition, you should be able to calculate irregular areas such as plots of land, using different methods.

P5 *produce clear and accurate answers to a variety of simple 2D trigonometric problems*

Trigonometric problems are based around triangles. Therefore you need to be able to calculate angles and distances involving right-angled and non right-angled triangles.

P6 *produce clear and accurate answers to a variety of simple geometric problems*

Geometric problems will involve trigonometry such as Pythagoras. However, the emphasis of this section will be on solving problems involving circles and arcs. This is particularly relevant when you need to draw curves for road alignments, building arches and so on.

P7 *describe and illustrate the use of statistics in the construction industry*

You need to be able to collect and present data using a variety of different methods, including bar charts, pie charts and pictograms. In addition, you need to be able to describe the advantages and disadvantages of using different types of 'averages', such as mean, mode and median.

M1 *select and apply a variety of algebraic methods to solve linear, quadratic and simultaneous linear and quadratic equations*

You need to be able to solve problems involving linear, quadratic and simultaneous equations. These will be similar to the problems in P3. However, instead of drawing graphs you will need to solve the equations mathematically using formulae. Simultaneous equations involve two equations with two unknowns.

85

You will also need to extend the skills gained in P2 by simplifying equations that contain brackets. Since this is a Merit criterion, you are expected to solve problems with minimal help from your tutor.

M2 *extract data, select and apply appropriate algebraic methods to find lengths, angles, areas and volumes for one 2D and one 3D complex construction industry related problem*

This is your chance to solve more complex problems by combining methods learnt in P4, P5 and P6. The majority of problems will be centred on the design and setting out of buildings and roads. You are expected to solve problems with minimal help from your tutor. Don't forget that all solutions should be set out clearly and include the proper units.

M3 *use standard deviation techniques to compare the quality of manufactured products used in the construction industry*

Manufactured products, such as steel, concrete and bricks, are made to a certain size or strength. For P7 you could have calculated the 'mean' value of a property of the material. However, this only has a limited value in terms of comparing different batches. In order to assess the quality of the material, it is essential to be able to calculate the standard deviation. You will be expected to calculate the standard deviation with minimal help from your tutor.

D1 *independently undertake checks on calculations using relevant alternative mathematical methods and make appropriate judgments on the outcome*

To satisfy D1 you need to demonstrate two things. First, you need to be able to check the results of problems you have solved, either by using a different method or by using a spreadsheet application, such as Microsoft Excel. Second, you need to be able to provide suitable conclusions to the problem. For example, you need to be able to use your calculations to assess whether the concrete used in building foundations will be strong enough.

It is expected that you will be able to work independently and that your work is presented to a high standard.

D2 *independently demonstrate an understanding of the limitations of certain solutions in terms of accuracy, approximations and rounding errors*

To achieve D2 you need to be able to evaluate the accuracy of your answers. You need to assess the accuracy of the original measurements. For example, if you calculate an area from a map, you need to ask yourself, 'How accurate is the map?' You need to assess how many decimal places to use in your calculations. For example, what difference does it make to the calculation of the volume of a cylinder if π is used to 2 rather than 9 decimal places?

Finally, you need to assess whether your final answers are realistic. For example, when dealing with materials, you will also need to find out how they are sold. You certainly won't be able to buy half a tile, and you will probably have to buy tiles in packs, so your final answers need to be appropriate.

Again, it is expected that you will be able to work independently and that your work is presented to a high standard.

ACTIVITY 1

USING A SCIENTIFIC CALCULATOR TO PERFORM CALCULATIONS

The invention of the scientific calculator has certainly taken the drudgery out of performing calculations. However, it is so easy to push the wrong button and the final result could end up 'rubbish'. For example, you might calculate the area of a window to be 4.5 mm^2 – this is a bit small, unless you are an ant! So remember:

garbage in = garbage out

Follow this general advice about using a scientific calculator.

1. Keep the calculator instruction booklet
You will be using a lot of different function keys, therefore you will need to know how to use them. If necessary, buy a new calculator – it's a worthwhile investment.

2. Learn how to store and recall numbers
Most calculators have more than one memory. This allows you to save intermediate calculations without re-entering. More importantly it allows you to save **all** the significant figures or decimal places, so errors due to rounding up too early are avoided.

3. Learn how to use the edit keys such as delete and insert
Modern calculators allow you to see the formula entered into the display. This allows you to correct the figures entered before pressing the '=' key. Using the scroll keys, you can edit the figures to perform repetitive tasks such as calculating percentages or coordinates.

4. Learn how to use brackets (always in pairs)
Use brackets to perform intermediate calculations. These should be entered as per the formula. However, don't be afraid to add your own brackets. For example, the fraction $\frac{5}{6}$ could be entered as 0.83333. An alternative way would be to enter (5 ÷ 6). This has the advantage of avoiding rounding errors.

5. Learn how to 'fix' the number of decimal places
Most calculators display answers in 'standard form', which is a good way to handle very large and very small numbers. However, this can be confusing when converting say millimetres to metres.

For example, to convert 5.6 millimetres to metres we divide by 1000.

In standard form, the answer would appear as 5.6×10^{-03}. Some scientific calculators display this as 5.6^{-03} or $5.6E10^{-03}$. This means that we need to move the decimal point three times to the left to get the result 0.0056 m.

If we 'fix' the number of decimal places to 5, the answer would be displayed as 0.00560. This is a lot easier to read. The other advantage is that in 'fix' mode the result is automatically rounded up. When calculating the price of materials it is good to fix the display to two decimal places.

6. Always independently check your final answers
It is always important to double check your final answers. This can be done manually by rounding up using simple figures.

For example, if a room measures 5.3 m by 3.9 m it has an area of 20.67 m^2 area. This calculation can be checked manually by using rounded approximations of the room measurements. In this case, round the measurements to 5 m and 4 m respectively; it is simple to calculate that $5 \times 4 = 20$ m^2 which shows that the answer 20.67 m^2 contains no major mistake.

Worked example

One of the best ways to check that a building has been set out correctly is to measure the diagonals. To calculate the diagonal of a building measuring 5.75 m by 8.15 m we use the following formula: $c = \sqrt{a^2 + b^2}$ where c is the diagonal and a and b are the sides of the building. In this case $c = \sqrt{5.75^2 + 8.15^2}$.

Most modern calculators allow you to enter formulae as they are written so we could key in the following steps: $\sqrt{5.75^2 + 8.15^2} = 72.1725$. But this answer is **wrong** – why ?

Firstly, do a manual check. Rounding up the dimensions, we have $6^2 + 8^2 = 36 + 64 = 100$ and the square root of 100 is 10. So the diagonal should be approximately 10 m.

Let's try again by adding some brackets: $\sqrt{(5.75^2 + 8.15^2)} = 9.9742$. Now this is **correct**. Can you see how the brackets round the fraction make all the difference?

So to check the setting out of the building the diagonal measurement should be **9.97 m**.

Incidentally, the final answer is quoted to two decimal places because the sides of the building were only measured to two decimal places.

Tasks

Calculate the following to 3 decimal places.

1. $\dfrac{12 + 14 + 15}{2 \times 5}$

2. $\dfrac{13.589}{9 - 6.3}$

3. $\dfrac{13.5 - (5.7 \times 3)}{7}$

4. $\dfrac{(12.7 - 14.2) \times (18.1 - 2025)}{\sqrt{56}}$

5. $\sqrt{25.5^2 - 9.5^2}$

6. $\frac{1}{3} \times \pi \times 10(5.5^2 + 2.2^2 + (5.5 \times 2.2))$

ACTIVITY 2

USING FORMULAE AND REARRANGING EQUATIONS

Construction mathematics is simply about using formulae to solve problems. It is essential that, to avoid mistakes, you follow these steps.

1. Decide on the method to be used
Remember that there will be a number of different ways to solve a problem. It depends on what information you are given to start with. Drawing a simple sketch will often help clear the mind.

For example, suppose you want to calculate the amount of tarmac needed to resurface a car park. This will involve areas and volumes. To calculate the area will depend on the car park shape. Has it regular or irregular sides? The shape will determine which formula you need to use.

2. Decide on the formulae to be used
These are simply taken from the textbook or formulae sheet. Remember that letters are used in formulae to represent numbers. It is a lot easier to write π instead of 3.142 159 265 4 every time. Because letters are used in many formulae, avoid using 'x' for multiplication, so 'ab' would actually mean 'a times b' ($a \times b$). Think of it as the maths version of texting.

3. If necessary, rearrange the formulae
This is the stage when you can simplify or manipulate the formulae. This is so important that it must practised until it becomes second nature. The rules for manipulating or transposing formulae are given below.

4. Substitute the values into the formulae
Rewrite the formulae with the actual figures.

5. Calculate the answer
Work out the answer. Don't forget the advice given in Activity 1.

6. Check the answer

Carry out a rough check and ask yourself the question, 'Does the answer makes sense?'

Make sure that you add the correct units and round up to the appropriate number of decimal places or significant figures.

The rules for rearranging equations (also known as transposing)

Every equation is separated by '=', ie the left side **equals** the right side. Look at these following equations:

a) $y + 6 = 10$ What is 'y'? Instinctively we have worked out that $y = 4$. In our heads we calculated $10 - 6 = 4$.

b) $y - 6 = 10$ $y = 16$, that is $10 + 6$

c) $6y = 10$ $y = 1.667$, that is $\left[\frac{10}{6}\right]$

d) $\frac{y}{6} = 10$ $y = 60$, that is 6×10

Can you see a pattern? When you move numbers or letters (called terms) from the left side to the right side and vice versa, you do the opposite to the function. In other words:

$+$	becomes	$-$	$\times^2$	becomes	$\sqrt{}$	
$-$	becomes	$+$	$\sqrt{}$	becomes	$\times^2$	
$\times$	becomes	$\div$	$\sin$	becomes	$\sin^{-1}$	
$\div$	becomes	$\times$	$\sin^{-1}$	becomes	$\sin$	

Worked example

In Activity 1 we converted 53° Fahrenheit to Celsius. We can transpose the formula to convert Celsius to Fahrenheit as follows.

Start with the original formula: $C = \frac{5}{9}(F - 32)$

First, move the 9 to the other side of the equation. As 9 is dividing, so it becomes multiply: $9C = 5(F - 32)$

Then, move the 5. This is multiplying, so it divides when we move it to the other side, and we can also now ignore the brackets: $\frac{9C}{5} = F - 32$

Minus 32 becomes plus: $\frac{9C}{5} + 32 = F$

This gives the final formula to convert Celsius to Fahrenheit – but does it work? From Activity 1, we know that if $C = 11.7$ then we should get 53°F.

So: $\frac{9 \times 11.7}{5} + 32 = 53.060$

The apparent error of 0.060 is due to rounding up the original value previously calculated. If we had used 11.66667 then the answer would have been 53.000 (to 3 decimal places). This is a good example to show that if you don't use enough decimal places you can end up with wrong answers!

Tasks

Transpose the following equations.

1. $A = \pi r^2$ What is r?

2. $c^2 = a^2 + b^2$ What is b?

3. $0 = 2x - 4y$ What is y?

4. $\frac{H}{t} = \frac{kA(\theta_1 - \theta_2)}{d}$ What is k?

5. $a^2 = b^2 + c^2 - 2bc \cos A$ What is $\cos A$?

6. $\frac{a}{\sin A} = \frac{b}{\sin B}$ What is B?

ACTIVITY 3

USING GRAPHICAL METHODS TO SOLVE EQUATIONS

The most common graphical method used to solve equations is a graph. Linear equations such as $y = 2x + 4$ produce a straight line, whereas a quadratic equation (which has an x^2 term, such as $y = 2x^2 + 3$) would be a symmetrical curve.

General advice about plotting graphs can be summarised as follows.

1. Use graph paper
Pretty obvious really. Graphs produced by software applications such as Microsoft Excel are of limited value because it is difficult to interpret intermediate values.

2. Choose the largest scale possible
The x-axis and y-axis can have different scales. However, don't complicate things by using odd units. For example, suppose the x-axis was drawn to a scale such that each 20 mm represented 15°C. This means that 1 mm would represent 0.75°C. It would be better to draw the graph using a slightly smaller scale so that each mm represent 1°C.

3. Label the axes
Always add a title to the graph and label the x-axis and y-axis, including the appropriate units.

4. Plotting equations
Any equation can be plotted by substituting different values for x and calculating the corresponding values for y. The coordinates can then be plotted. Using this method you can plot any equation. Consider, for example, the linear equation $y = 2x + 4$.

x	Calculation	y
-5	$(2 \times -5) + 4$	-6
-4	$(2 \times -4) + 4$	-4
-3	$(2 \times -3) + 4$	-2
-2	$(2 \times -2) + 4$	0
-1	$(2 \times -1) + 4$	2
0	$(2 \times 0) + 4$	4
1	$(2 \times 1) + 4$	6
2	$(2 \times 2) + 4$	8
3	$(2 \times 3) + 4$	10
4	$(2 \times 4) + 4$	12
5	$(2 \times 5) + 4$	14

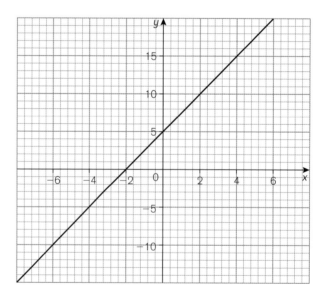

Worked example

Problem 1

A site engineer working in the Middle East uses a 30 m steel tape to set out a new factory. However, the steel tape will expand and contract due to changes in temperature. The temperatures on site vary from $-2°C$ in the early morning to 45°C in the afternoon. Produce a graph to correct the tape measurements.

Assume the steel tape expands and contracts at a rate of 0.000 01125 m/deg C and that the maximum distance measured using the tape would be 40 m. Steel tapes are generally calibrated to measure 30 m at 20°C.

1. Method
    We need to plot a graph to show the necessary corrections that must be made when measuring 40 m at any temperature between the minimum and maximum possible temperatures. Take these to be −5°C and 50°C respectively, just in case the temperatures get even colder or hotter. The graph will have the y-axis as air temperature and the x-axis as the corrected distance.

2. Formulae
    Correction = measured distance × rate of expansion × temperature difference from the standard

    Correct distance = measured distance ± correction

    (add the correction if the tape contracts and subtract the correction if the tape expands)

3. Calculations
    Calculate the correction and the correct distance at the maximum and minimum temperatures.

    At 50°C,
    Correction = $(40 \times 0.000\,01125 \times (20 - 50))$
    $= \mathbf{-0.0135\,m}$

    Correct distance = 40 − 0.0135
    $= \mathbf{39.987\,m}$ (to 3 decimal places)

    Because the tape will have expanded, when working at 50°C the site engineer must set out 39.987 m in order to set out 40 m correctly.

    At −5°C,
    Correction = $(40 \times 0.000\,01125 \times (20 - (-5)))$
    $= \mathbf{+0.01125\,m}$

    Correct distance = 40 + 0.01125
    $= \mathbf{40.011\,m}$ (to 3 decimal places)

    Because the tape will have contracted, when working at −5°C the site engineer must set out 40.011 m in order to set out the 40 m correctly.

    By plotting these two coordinates, we can produce a graph that would allow the engineer to read off the correct distance to measure to set out 40 m when working at any given temperature.

x-axis (distance m)	y-axis (temperature °C)
40.011	−5
39.987	+50

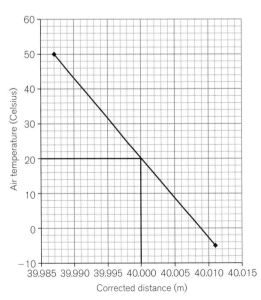

4. Check
    If the graph is drawn correctly then the correction at 40 m should be zero at 20°C, ie the corrected distance should be 40 m exactly.

Problem 2

A rectangular piece of land of area 75 m² is required to be sold. The planning authority requires the length of the land to be 3 m longer than the width. Calculate the dimensions of the rectangle.

1. **Method**

   Start by making the width x, so the length must be $x + 3$. We know the area is 75 m², so that $x(x + 3)\text{m}^2 = 75\,\text{m}^2$.

   This can be rearranged to give the equation $x^2 + 3x - 75 = 0$. This is a quadratic equation so when plotted it will be a curve. We need to plot $y = x^2 + 3x - 75$. The values of x will have to be positive since land cannot have negative dimensions. We need to calculate coordinates by giving x a range of values. Once plotted the required value of x will be where the graph crosses the x-axis, ie when y = zero.

2. **Formula**

   Substitute a range of values for x, say from 5 to 10 (trial and error) into the formula $y = x^2 + 3x - 75$.

3. **Calculations**

x	Calculation	y
5	$5^2 + (3 \times 5) - 75$	-35
6	$6^2 + (3 \times 6) - 75$	-21
7	$7^2 + (3 \times 7) - 75$	25
8	$8^2 + (3 \times 8) - 75$	13
9	$9^2 + (3 \times 9) - 75$	33
10	$10^2 + (3 \times 10) - 75$	55

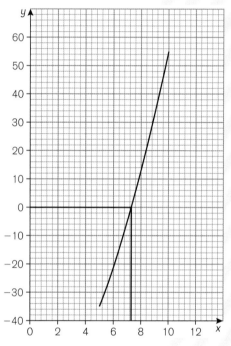

The graph crosses the x-axis at approximately at 7.3 m. Therefore the dimensions of the rectangular plot of land should be 7.3 m by 10.3 m.

4. **Check**

   Calculate the area of the land: $7.3 \times 10.3 = 75.190\,\text{m}^2$. This is just over the required area of 75 m². The only way to get a more precise value would be to use a larger scale on the x-axis or solve the equation by analytical methods.

Tasks

1. A site engineer working in Australia uses a 30 m steel tape to set out a new road. The temperatures vary from −2°C in the morning to 45°C in the afternoon. Produce a graph to correct the tape measurements.

   Assume the steel tape expands and contracts at a rate of 0.000 01125 m/deg C and that the maximum distance measured using the tape would be 50 m. Steel tapes are generally calibrated to measure 30 m at 20°C.

2. Draw a graph to convert miles to kilometres and vice versa. Use the graph to convert 35 miles to kilometres and 80 kilometres to miles. Note that 1 mile is 1.609 344 km.

3. A building has a dome roof. This is designed in the form of a segment of a sphere. Assuming the radius of the dome is 12 m and its volume is 500 m³, the height of the dome can be expressed as:

$$h^3 - 36h^2 + (1500 \div \pi) = 0$$

Note: this is called a cubic equation because of the h^3 term. It is a lot easier to solve this type of equation by drawing a graph.

Plot a graph of the equation to find the value of h to the nearest 50 mm, given that its value lies between 3 m and 4 m.

Use this formula to check your answer.

Volume $= \dfrac{\pi h^2}{3}(3r - h)$

Where r is the radius and h is the height.

4. A bricklayer wants to build a brick arch as shown in the sketch below. The span (S) is 3 m and the radius (R) is 6 m.

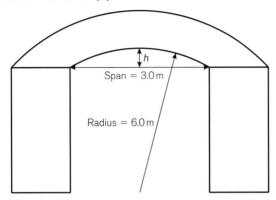

The relationship between the span, the radius and the height of the arch (h) can be expressed as:

$$8Rh - S^2 - 4h^2 = 0$$

Plot a graph of the above equation to find the value of h to the nearest 10 mm, given that its value lies between 0.15 m and 0.25 m.

Use this formula to check your answer.

$$R = \dfrac{S^2 + 4h^2}{8h}$$

ACTIVITY 4

CALCULATING AREAS AND VOLUMES

The need to calculate areas and volumes is very common in construction. These are used mainly for calculating quantities of materials required, for heat loss calculations, and for issues involved in buying and selling land.

Here is some general advice on calculating areas and volumes.

1. Use the correct units

When calculating areas and volumes, only use metres or millimetres. Do not use centimetres.

Length	metres (m)	
Area	square metres (m²)	[1 hectare = 10 000 m²]
Volume	cubic metres (m³)	[1 litre = 0.001 m³]

1 m = 1000 mm	[mm to m	divide by 1000]
1 m² = 1 000 000 mm²	[mm² to m²	divide by 1 000 000]
1 m³ = 1 000 000 000 mm³	[mm³ to m³	divide by 1 000 000 000]

Remember that areas have 'square' units and volumes have 'cube' units.

2. Formulae sheet

Design your own formulae sheet. Unless you have access to an 'equation editor' on your computer, write them out by hand.

3. Regular areas and volumes

Regular area problems can usually be broken down into simple shapes, such as squares, rectangles, triangles and circles. Here are a couple of useful formulae that can be used to calculate the area of a trapezium and a triangle respectively.

Trapezium

$$\text{Area} = \frac{d(a + b)}{2}$$

Where a and b are the lengths of the parallel sides and d is the distance between them.

Heron's formula

This can be used to calculate the area of a triangle if you know the lengths of its three sides. If its sides are lengths a, b and c, then:

$$\text{Area} = \sqrt{s(s - a)(s - b)(s - c)} \text{ where } s = \left(\frac{a + b + c}{2}\right)$$

Volumes

The golden rule when calculating volumes for objects having a uniform cross-sectional area is to calculate the cross-sectional area **first** and then multiply this answer by the object's length (or height) – it's as simple as that!

The diagram below shows a concrete pipe. The best method to calculate the volume of concrete required is to calculate the cross-sectional area first (the shaded part of the end) and then multiply this answer by the pipe's length.

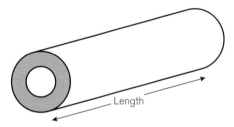

Length

4. Irregular areas and volumes

You will never get the 'true area' or 'true volume' because, by their very nature, irregular features are open to interpretation. The results from Simpson's rule and the trapezoidal rule will be different. It's up to you to decide which is the more accurate! As a rule of thumb, Simpson's rule gives a more accurate result if the perimeter is more irregular and curved. Notice how the Simpson's and trapezoidal rules have similar formulae for both areas and volumes.

Simpson's rule

Area $= \frac{d}{3}$ [first ordinate + last ordinate + 2($\sum$ remaining odd ordinates) + 4($\sum$ remaining even ordinates)]

Where d is the distance between ordinates and $\sum$ means 'the total or the sum'. Note that there **must** be an odd number of ordinates, ie an even number of strips. An example of how to use Simpson's rule to calculate the area of an irregular feature is given in one of the worked examples below.

Volume $= \frac{d}{3}$ [first area + last area + 2($\sum$ odd areas) + 4($\sum$ even areas)]

Where d is the distance between cross-sectional areas. Note that there must be an odd number of areas.

Trapezoidal rule

Area $= \frac{d}{2}$ [first ordinate + last ordinate + 2($\sum$ remaining ordinates)]

Where d is the distance between ordinates.

Volume $= \frac{d}{2}$ [first area + last area + 2($\sum$ remaining areas)]

Where d is the distance between cross-sectional areas.

Worked example

Problem 1

A builder has been asked to construct a simple flat-roof garage. The details of the garage are shown below. The builder wants to calculate the number of bricks required.

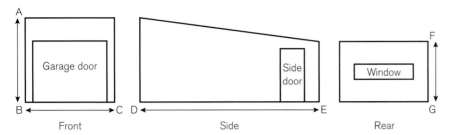

The size of the front is 3.0 m (AB) by 2.8 m (BC) and the length (DE) is 5 m. The rear is lower than the front; FG is 2.5 m. The garage door is to be 2 m by 2 m, the side door 0.8 m by 2 m and the rear window 1.2 m by 0.6 m.

1. Method

    We need to calculate the area of the four walls. The front and rear are rectangles and the two sides are trapeziums. The doors and windows are all rectangles.

    We need to calculate the total (gross) area and then subtract spaces for the doors and windows. This will give us the area for the bricks; we can then work out the number of bricks required.

    An allowance for wastage should be included, say 5 per cent, and we usually assume 60 bricks per square metre.

2. Formulae

    Area of a rectangle = length × width

    Area of a trapezium $= \dfrac{d(a + b)}{2}$

    Where a and b are the lengths of the parallel sides and d is the distance between them.

3. Calculations

    Producing a table improves the presentation.

Element	Length (m)	Height (m)	a (m)	b (m)	d (m)	Calculation	Area (m²)	
Front	2.8	3.0				2.8 × 3.0	8.40	
Rear	2.8	2.5				2.8 × 2.5	7.00	
Side 1			3.0	2.5	5.0	5(3.0 + 2.5) ÷ 2	13.75	
Side 2			3.0	2.5	5.0	5(3.0 + 2.5) ÷ 2	13.75	
						Total (gross)	42.90	[1]
	Length (m)	Height (m)						
Garage door	2.0	2.0				2.0 × 2.0	4.00	
Side door	0.8	2.0				0.8 × 2.0	1.60	
Window	1.2	0.6				1.2 × 0.6	0.72	
						Sub total	6.32	[2]
						Area (net)	36.58	[1] − [2]

So the area required for the bricks is 36.58 m². There are 60 bricks per square metre, so the number of bricks required is:

36.58 × 60 = 2194.8 bricks

Add 5 per cent for wastage $= 2194.8 + \left(2194.8 \times \frac{5}{100}\right)$

$= 2304.54$ bricks

Although we would normally round up to 2305 bricks, realistically you would ask for 2300. It all depends on how good the bricklayers are!

4. **Check**

   A rough check on the area would be to calculate the perimeter and multiply by the average height.

   Area $= 2.75(2.8 + 5 + 5 + 2.8) = 2.75 \times 15.6 = 42.9\,m^2$

   Subtract the garage door which is the biggest opening ($4\,m^2$).

   The net area is approximately $39\,m^2$ ($43 - 4$), so the number of bricks needed would be $39 \times 60 = 2340$.

Problem 2

The local authority wishes to sell one of its allotment gardens to a housing developer for £200 per square metre. The plot is shown below. Calculate the area of the land to be sold.

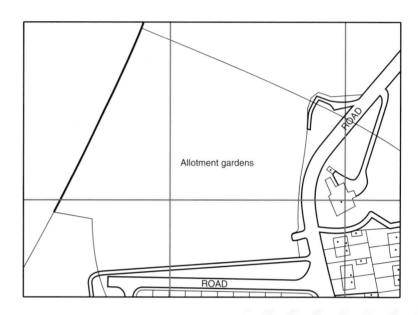

1. **Method**

   The area has irregular boundaries, so we can use Simpson's rule. We need to divide the area into parallel lines (ordinates). To apply Simpson's rule, there must be an odd number of ordinates. The length of each ordinate should be scaled in metres. Note that the scale is 1:1250.

2. **Formulae**

   Area $= \frac{d}{3}$ [first ordinate + last ordinate + 2($\sum$ odd ordinates) + 4($\sum$ even ordinates)]

   Where d is the distance between ordinates. Note that there **must** be an odd number of ordinates.

3. **Calculations**

   The first step is to mark the baseline points shown in circles. The distance apart between points is set at 10 m for convenience. The ordinates (dotted) are drawn with a set square at right angles to the baseline. The length of each ordinate is measured and recorded as follows.

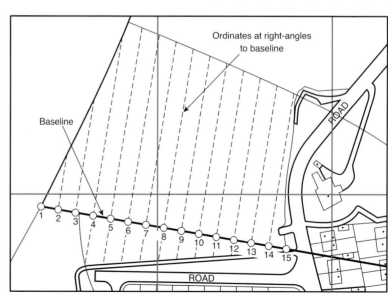

Ordinate	Length	Comment
1	0.0	First
2	35.0	Even
3	67.5	Odd
4	127.5	Even
5	133.0	Odd
6	128.5	Even
7	124.0	Odd
8	119.0	Even
9	114.5	Odd
10	109.0	Even
11	104.5	Odd
12	99.5	Even
13	94.5	Odd
14	89.0	Even
15	0.0	Last

To apply Simpson's rule we need to sum the odd ordinates and then sum the even ordinates.

The sum of the odd ordinates = 638.0
The sum of the even ordinates = 707.5

So apply the rule:

Area $= \frac{10}{3}[0 + 0 + (2 \times 638.0) + (4 \times 707.5)]$
$= 13\,686.667$
$= \textbf{13\,687 m}^2$

4. Check
    Using the trapezoidal rule the area would be:

    Area $= \frac{10}{2}[0 + 0 + 2 \times (638.0 + 707.5)]$
    $= 13\,455.000$
    $= \textbf{13\,455 m}^2$

    This is a difference of 231.667 m^2, which amounts to a price difference of roughly £46 333 at £200 per square metre. The housing developer would want to use the trapezoidal rule because this gives a smaller area and therefore would result in a cheaper price for the land.

Tasks

1. A new car park is to be set out and concreted. Using the measurements shown in the diagram below, calculate the area of the car park and the volume of concrete required, assuming it is to be laid with a thickness of 150 mm.

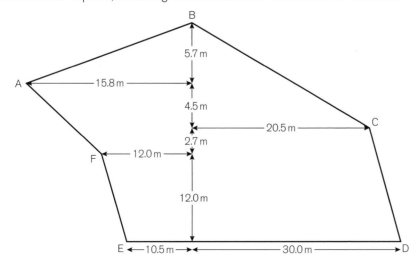

2. A new roof is to be constructed in the form of a frustum of a rectangular pyramid, as shown in the diagram below. The base of the roof is 6 m by 4 m. The top of the roof is 4 m by 2 m. The perpendicular height is 2.828 m (h) and the slant length is 3 m (s). Calculate: **a)** the volume of the roof space, and **b)** the total surface area of the roof excluding the base.

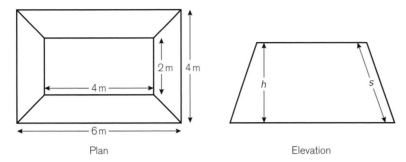

Plan   Elevation

3. Calculate the cross-sectional area of a river crossing shown below.

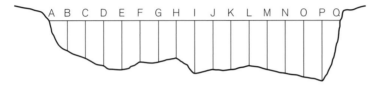

Depths were taken every 2 m as follows.

A	B	C	D	E	F	G	H	I	J	K	L	M	N	O	P	Q
0.0	0.8	1.0	1.2	1.3	1.1	1.1	1.0	1.4	1.3	1.3	1.2	1.3	1.4	1.5	1.6	0

4. Limestone taken from a quarry has been stockpiled and surveyed. A contoured plan has been produced and cross-sectional areas calculated below. Calculate the amount of material that has been excavated from a quarry.

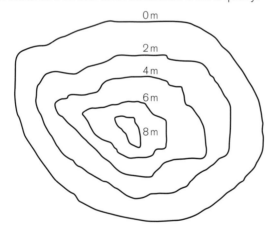

Section	Contour (m)	Area (m²)
1	0	1485
2	2	736
3	4	300
4	6	91
5	8	15

ACTIVITY 5

USING TRIGONOMETRY TO SOLVE PRACTICAL PROBLEMS

Trigonometry comes from the Greek words *trigonon* (triangle) and *metria* (measure). So trigonometry is about using triangles to help solve problems. It is used a lot in construction, especially in surveying and building design for the measurement of angles and distances. The skill is to use triangles to solve real-life problems, such as to calculate the height of a building.

1. Using angles

In the UK, angles are divided into *degrees*, *minutes* and *seconds*.

1 degree = 60 minutes; 1 minute = 60 seconds

In a full circle there are 360°. A right angle would be written as follows: 90° 0′ 00″.

Incidentally, some European countries use gons (g) instead of degrees. In this system there are 400 g in a full circle, hence a right angle is 100 g. So be careful if you work overseas.

It is essential that you know how to set your calculator to 'degree' mode and how to enter angles into the calculator. Check your calculator's instructions, and do the same if using a software program such as Microsoft Excel.

2. Types of triangle

There are only two types of triangle, right-angled and non-right-angled. There are only six elements to solve – three angles and three sides.

You cannot solve a triangle unless you know a minimum of three elements including at least one side.

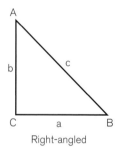
Right-angled

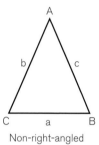
Non-right-angled

Labelling triangles is critical. We use **upper case** letters, such as A, B and C, for angles, and **lower case** letters, such as a, b and c, for sides.

In any triangle, $A + B + C = 180° \ 00' \ 00''$.

3. Formula for a right-angled triangle

In these rules, the longest side is always labelled c. This side is called the hypotenuse.

Pythagoras rule

$c^2 = a^2 + b^2$ ie $c = \sqrt{(a^2 + b^2)}$

Trigonometrical ratios

$\sin A = \dfrac{a}{c}$ or $\dfrac{\text{opposite side}}{\text{hypotenuse}}$

$\cos A = \dfrac{b}{c}$ or $\dfrac{\text{adjacent side}}{\text{hypotenuse}}$

$\tan A = \dfrac{a}{b}$ or $\dfrac{\text{opposite side}}{\text{adjacent side}}$

4. Formula for a non-right-angled triangle

You cannot use Pythagoras in a non-right-angled triangle.

Sine rule

$$\frac{a}{\sin A} = \frac{b}{\sin B} = \frac{c}{\sin C}$$

Cosine rule

$a^2 = b^2 + c^2 - (2bc \cos A)$ or $\cos A = \dfrac{b^2 + c^2 - a^2}{2bc}$

Worked example

A new pier is to be constructed at a small seaside town. Three reference beacons R1, R2, and R3 have been set out in the form of a right-angled triangle as shown below. The length of the pier is to be 200 m from R2 and a barge is to be use to drive piles into the seabed. The captain aligns the barge with R1 and R2 and measures the angle 17° 20′ between R2 and R3 using a sextant. Calculate the distance from R2 to the barge and hence the distance required to the end of the pier. Then calculate the sextant angle that the captain should read when the barge is at the end of the pier.

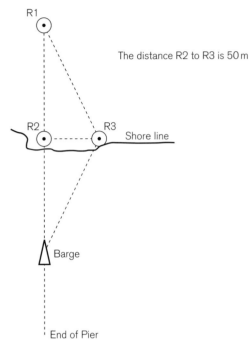

The distance R2 to R3 is 50 m

Shore line

Barge

End of Pier

1. **Method**

   This problem involves two right-angled triangles. Relabelling the triangle, we have:

   Angle A is 17° 20′
   Opposite side (a) is 50 m
   Adjacent side (c) is to be calculated

   Therefore the tan ratio should be used to calculate the current position of the barge. Subtracting this distance from 200 m will give the distance to the end of the pier.

   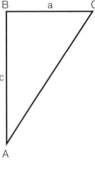

   To calculate the angle between R2 and R3 when the barge is at the end of the pier we have another right-angled triangle:

   Angle A is unknown
   Opposite side (a) is 50 m
   Adjacent side (c) is 200 m

   Again, the tan ratio should be used to calculate the required sextant angle.

2. **Formulae**

   $$\tan A = \frac{a}{c} \text{ and } A = \tan^{-1}\left(\frac{a}{c}\right)$$

   Transposing, this gives:

   $$c = \frac{a}{\tan A}$$

3. Calculations

$$c = \frac{50}{\tan 17° 20'} = 160.203 \, m$$

Distance to the end of the pier is $200 - 160.203 = 39.797 \, m$

In other words, the barge is approximately 40 m from the end of the pier.

To calculate the angle at the end of the pier:

$$A = \tan^{-1}\left(\frac{50}{200}\right) = 14° \, 02' \, 10''$$

Given the accuracy of the sextant, the captain of the barge would be looking for 14° 00′ to the nearest 10′ of arc.

4. Check

A simple independent check would be to calculate the length AC, distance from the barge to R3.

$$\sin A = \frac{a}{b} \quad \text{and} \quad b = \frac{a}{\sin A} \qquad b = \frac{50}{\sin 17° 20'} = 167.825 \, m$$

This shows that hypotenuse is longer, as expected, than the adjacent length AB.

Tasks

1. This diagram shows a frame structure for a roof.

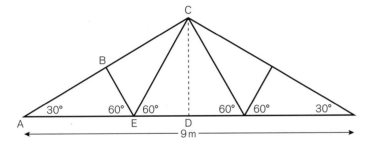

Assuming that this is a symmetrical roof, ie $AD = 4.5 \, m$, calculate the height of the roof CD and the following lengths: AB, BC, BE, CE, EA.

2. A house owner has built a garage and wants to know whether a neighbour's tree at the end of the garden would hit the garage if it were to fall over. A surveyor standing on top of the garage has measured the slope distance to the base of the tree and has taken vertical angles to the base and top of the tree. Would the tree hit the garage?

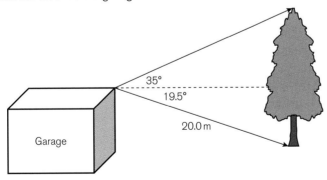

ACTIVITY 6

USING GEOMETRY TO SOLVE PRACTICAL PROBLEMS

Geometric problems often involve circles and angles, so there is a strong link with trigonometry. However, in this section we will investigate problems involved with circles and using radians. General advice about using circles can be summarised as follows.

101

1. Basic terminology

You should know the difference between these common terms.

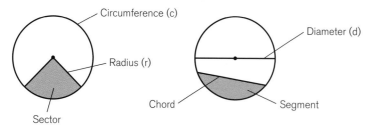

The ratio between the circumference and the diameter is called π (pi). This means that:

$$\pi = \frac{\text{circumference}}{\text{diameter}} \text{ so that } c = \pi d \qquad \text{or } 2\pi r$$

Since $d = 2r$, we can also write this equation as $d = 2\pi r$

The area of a circle (A) is give by:

$$A = \pi r^2 \text{ or } \frac{\pi d^2}{4}$$

2. Radians

Angles measured at the centre of a circle are expressed in degrees, minutes and seconds; in a full circle there are 360°.

An alternative way to express angles is by measuring the length of arc. In a full circle the arc would be $2\pi r$ in length, ie the length of the circumference. We define a radian by assuming the radius is 1. In a full circle there are 2π radians. This means that:

$$1° = \frac{\pi}{180} \text{ radians}$$

To convert degrees to radians, we multiple the angle by $\frac{\pi}{180}$

3. Formulae involving radians

The arc length $= r \times \theta$ radians   **or** $r\theta° \times \frac{\pi}{180}$

Note that symbol θ is pronounced 'theta'.

The area of a sector $= \frac{r^2}{2} \times \theta$ radians   **or** $r^2\left(\frac{\pi\theta°}{360}\right)$

The area of a segment $= r^2\left(\frac{\pi\theta°}{360} - \frac{1}{2}\sin\theta°\right)$

Worked example

A highways engineer has designed a road for a new housing estate. The centre line of the road is shown below. The site engineers need to be able to calculate the arc length and chord length between the two tangent points (TP1 and TP2) so that they can set out the road on the ground.

1. **Method**
   To calculate the arc length is straightforward because we have the radius and the angle. However, the angle must be converted into radians. We have choice when calculating the chord length. The chord is part of a non-right-angled triangle, therefore the cosine rule would be most appropriate (there is not enough information to use the sine rule).

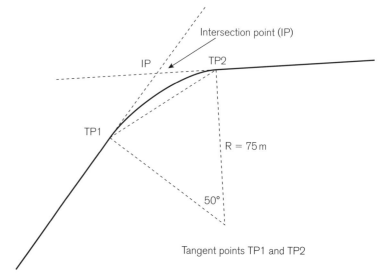

Tangent points TP1 and TP2

2. Formulae

    The arc length $= r\theta° \times \dfrac{\pi}{180}$

    The cosine rule is $a^2 = b^2 + c^2 - (2bc \cos A)$

3. Calculations

    Arc length $\dfrac{75 \times 50 \times \pi}{180} = 65.449\,847$

    The arc length is **65.450 m**.

    To calculate the chord length:

    a = the chord length, $b = 75\,\text{m}$, $c = 75\,\text{m}$ and $A = 50°$

    Using the cosine rule:

    $a^2 = 75^2 + 75^2 - (2 \times 75 \times 75 \times \cos 50°)$

    $a^2 = 4018.639\,391$

    $a = \sqrt{4018.639\,391}$

    $a = 63.392\,739$

    Chord length is **63.393 m**.

4. Check

    Firstly, the arc length should always be longer than the chord length. In this case the arc length is 2.057 m longer.

    The chord length can be checked a different way, by splitting the triangle into two right-angled triangles.

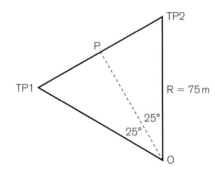

    The chord length is now twice the length P to TP2, so, using the formula:

    $\sin A = \dfrac{\text{opposite}}{\text{hypotenuse}}$

    The chord length $= 2 \times 75 \times \sin 25° = $ **63.393 m** (same result as in Point 3).

Tasks

1. The kerb line for a new road junction needs to be set out. Using the information below, calculate the radius of the circle (of which the curved section of the kerb forms a part) and arc length (TP1 to TP2).

TP1   IP

Chord length = 5.550 m

TP2

intersection angle $\theta = 125°$

103

2. In order to calculate the velocity of water in a drainage pipe it is necessary to calculate the hydraulic mean radius (HMR).

The HMR $= \dfrac{\text{cross sectional area of water}}{\text{wetted perimeter}}$

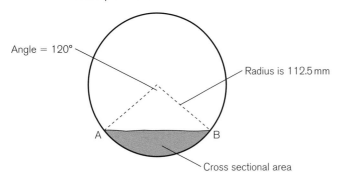

Angle = 120°

Radius is 112.5 mm

A   B

Cross sectional area

Calculate:
a) the cross-sectional area
b) the wetted perimeter (arc *AB*)
c) the hydraulic mean radius.

ACTIVITY 7

USING STATISTICS IN CONSTRUCTION

"There are three kinds of lies: lies, damned lies and statistics." This is a well-known saying attributed to the nineteenth century British Prime Minister Benjamin Disraeli. The implication is that you can use statistics to back up any argument. The quotation below is from the Health and Safety Executive.

> **Construction deaths down in 2004/2005 – but not a time to be complacent**
> Statistics released yesterday by the Health and Safety Executive (HSE) show the total number of fatal injuries to workers in construction for this year is 72, a small increase on 71 workers in 2003/04. However, due to a continued rise in employment, the fatal injury rate has fallen by 3% to 3.48 per 100,000 workers, continuing the downward trend of the past four years. This is the lowest level seen on record.
>
> *Source:* HSE press release

So the number of deaths has gone up *and* down – both are equally valid interpretations of the same statistic. Whenever you are presented with statistical analysis you must question how the information was both collected and presented.

General advice about statistics can be summarised as follows.

1. Presenting information
The most common ways of presenting information are by a table, a bar chart, a pie chart, a line graph and a symbolic chart or graph. Whenever you use any of these methods always make sure that you include a title and provide the source and date of the information.

For graphs:

• the scale on both axes should be regular
• axes should be clearly labelled and include proper units
• the 'areas' drawn should be proportional to the frequency.

2. Averages
An 'average' is a single number used to represent a whole group or data set. There are four methods: mode, median, arithmetic mean and geometric mean.

Mode

The mode is the most frequent item or number that occurs within a data set. It is simple to determine, but if all the data is different you could end up with no mode. Equally you could end up with two or more modes.

The mode would be useful when ordering stock. For example, many developers allow customers choose paint colours for their new houses, and builders would want to ensure that they had a stock of the colour that is chosen most often.

Median

This is the middle value of a set of data. It can be tedious to calculate if there is a large amount of data and you don't have software (such as Microsoft Word or Excel) to sort the data. The main advantage of the median is that it excludes extreme values.

An example of the use of the median would be when a client decides to award a building contract. Some companies will put in very high or very low prices. Either they don't really want the work or they are desperate for work – and the last thing a client wants is for the building company to go bankrupt in the middle of a contract. A sensible option might be to choose the median bid.

Arithmetic mean

This is the most popular average and has its own symbol $\bar{x}$. You can only calculate an arithmetic mean with numerical data.

$\bar{x} = \frac{\sum x}{n}$ where n is the number of items

$\bar{x} = \frac{\sum xf}{\sum f}$ where f is the frequency of items

The advantage of this method it that all the data are used. However, the result can be biased (or skewed) if there are abnormally high or low values in the data.

Geometric mean

In this case the data is multiplied and then the $\frac{1}{n}$th root is calculated.

Geometric mean $= (a_1 \times a_2 \times a_3 \times a_4 \times \ldots a_n)^{\frac{1}{n}}$

The main advantage of this type of average is that it compensates for abnormal values and so shows less bias than the arithmetic mean. Incidentally, you could not work out the geometric mean if the data contains a zero – think about it!

Worked example

During the construction of a new building a batch of concrete is produced. Eleven samples of the concrete have been tested for strength. Calculate the average strength of the concrete.

Sample	1	2	3	4	5	6	7	8	9	10	11
Strength N/mm²	36	35	36	39	65	37	66	35	38	35	35

1. Method

   To calculate and compare the four averages – the mode, the median and the arithmetic and geometric means.

2. Formulae

   Mode – most frequent value

   Median – put data in numerical order and find middle value

   Arithmetic mean $= \bar{x} = \frac{\sum x}{n}$

   Geometric mean $= (a_1 \times a_2 \times a_3 \times a_4 \times \ldots a_n)^{\frac{1}{n}}$

3. Calculations

Mode = **35 N/mm²**, by inspection of the data.

Median requires putting the values in order

Rank	1	2	3	4	5	6	7	8	9	10	11
Strength N/mm²	35	35	35	35	36	36	37	38	39	65	66

Middle value is **36 N/mm²**

Arithmetic mean $= \dfrac{35 \times 35 \times 35 \times 35 \times 36 \times 36 \times 37 \times 38 \times 39 \times 65 \times 66}{2} = $ **41.5 N/mm²**

Geometric mean $= (35 \times 35 \times 35 \times 35 \times 36 \times 36 \times 37 \times 38 \times 39 \times 65 \times 66)^{\frac{1}{11}} = $ **40.3 N/mm²**

4. Check

These are the four values (to 2 significant figures):

Mode   = **35 N/mm²**
Median   = **36 N/mm²**
Arithmetic mean = **42 N/mm²**
Geometric mean = **40 N/mm²**

Of course the problem now is to decide which of the four 'averages' best represents the whole batch of concrete – what do you think?

Tasks

1. This table shows the monthly earning for two types of construction workers.

Monthly earnings	Quantity surveyors	Bricklayers
East Midlands	£ 2355.95	£ 1585.21
Eastern	£ 3385.63	£ 1458.34
London	£ 3233.64	£ 2398.39
North East	£ 2594.54	£ 1635.01
North West	£ 2575.48	£ 1499.05
Scotland	£ 2362.88	£ 1337.10
South East	£ 3244.47	£ 1832.46
South West	£ 2568.99	£ 1510.74
Wales	£ 2051.99	£ 1645.40
West Midlands	£ 2732.66	£ 1887.88
Yorkshire and Humberside	£ 2877.72	£ 1700.39

Source: www.worksmart.org.uk/tools/paywizard.php

Calculate the median, arithmetic mean and geometric mean monthly earnings for quantity surveyors and bricklayers. Which method do you think best represents UK monthly earnings?

2. Using the data from question 1 compare the two careers by drawing a bar chart and a line diagram. What conclusions could you draw from the graphs?

3. Look at the following graphs and diagrams and suggest why they are misleading.

   a) Construction deaths

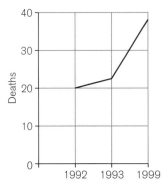

b) Sales

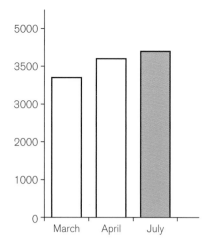

c) Use of IT on construction sites

England

Wales

ACTIVITY 8

HOW TO SOLVE LINEAR, QUADRATIC AND SIMULTANEOUS EQUATIONS

In Activity 3 we saw how to solve linear and quadratic equations using graphical methods. In this section we will look at alternative ways to solve these equations. Here is some general advice about solving simultaneous and quadratic equations.

1. The number of equations and unknowns

Remember that you can only solve an equation if the number of unknowns equals the number of equations.

For example, $y = 4x + 7$ can't be solved unless y or x is known. However, if you have two equations (called simultaneous equations) then both y and x can be solved. This is an example of a set of simultaneous equations:

$y = 2x + 6$
$y = 3x^2 + 4$

2. Substitution

Substitution is about converting two equations into one.

For example, if $y = 2x + 6$ and $y = 3x^2 + 4$ then $2x + 6 = 3x^2 + 4$ because both the left-hand side and the right-hand side equal y.

Rearranging we have $3x^2 - 2x - 2 = 0$. This is called a quadratic equation.

3. To solve a quadratic

A quadratic is one equation with one unknown and it must have a squared term, such as x^2.

Although there are different ways to solve a quadratic, the best way is to use this standard formula:

$$x = \frac{-b \pm \sqrt{b^2 - (4ac)}}{2a}$$

All quadratics will give you two values for x, hence the $\pm$ term in the equation. The letters a, b and c are called constants and are based on the standard form of writing quadratic equations, that is $ax^2 + bx + c = 0$

Note: the equation equals zero; 'a' links to the x^2 term; and 'b' links to the x term.

Therefore, using the example above, $3x^2 - 2x - 2 = 0$

$a = 3; b = -2; c = -2$

By substituting into the standard formula, the values of x are solved.

4. Brackets

You will need to know how to remove brackets from an equation.

Consider the following equation: $2(r + 9) = 26$

To remove the brackets the '2' multiplies both terms inside the brackets, so $2(r + 9) = 26$ becomes $2r + 18 = 26$ which is easier to solve.

It is possible to have equations with multiplying brackets such as $(2r + 4)(r - 5) = 0$. Each term in the first bracket multiplies the second bracket as follows:

$2r(r - 5) + 4(r - 5) = 0$

Multiplying out the brackets we have: $2r^2 - 10r + 4r - 20 = 0$

This can be further simplified to $2r^2 - 6r - 20 = 0$, which just happens to be a quadratic.

Worked example

Rainwater is stored in a 5 m high tank. The tank is made up of a cylinder below a circular dome (hemisphere) top. The height of the cylindrical section of the tank is h and the cylinder has a radius r. The tank has a total surface area of 138 m². It can be shown that:

$$h = 5 - r$$
$$3\pi r^2 + 2\pi rh = 138$$

Using these simultaneous equations, calculate the radius (r) of the water tank.

1. Method

   The value of h is expressed in terms of the radius r, therefore h can be substituted into the main equation.

   The r^2 term indicates that this is a quadratic therefore the equation must be rearranged into the form $ar^2 + br + c = 0$. Then r can be solved by using the quadratic formula.

2. Formulae

   The formula to solve a quadratic is: $r = \dfrac{-b \pm\sqrt{b^2 - (4ac)}}{2a}$

3. Calculations

   Substituting for h we have:

   $3\pi r^2 + 2\pi r(5 - r) = 138$

   $3\pi r^2 + 10\pi r - 2\pi r^2 = 138$ [removed the brackets]

   $\pi r^2 + 10\pi r - 138 = 0$ [rearranged the equation]

   $a = \pi, b = 10\pi, c = -138$ [remember π is just a number]

   $r = \dfrac{-10\pi \pm\sqrt{(10\pi)^2 - (4\pi \times -138)}}{2\pi}$

   $r = \dfrac{-10\pi \pm\sqrt{2721.119584}}{2\pi}$

   $r = 3.302214$ **or** $r = -13.302214$

   The radius is 3.302 m because you couldn't have a negative dimension for the storage tank.

4. Check
    Using the value of 3.302 m for the radius we can substitute it into the original equation.

    $h = 5 - 3.302$

    $h = 1.698$

    $3 \times \pi \times 3.302^2 + 2 \times \pi \times 1.698 = 137.988\,815$ [almost 138]

Tasks

1. An electric cable hanging between two pylons will sag. The shape of the cable is called a catenary. The length of the catenary is expressed as follows:

    $$L = \frac{8S^2}{3d} + d$$

    The length of the cable (L) is 60 m and the sag (S) is 1.750 m. Calculate the distance between the two pylons, ie the distance d.

2. A farmer has a field of 0.75 hectares and wants to sell part of the feild (as shown in the diagram below) to a builder. The farmer only needs 200 m of barbed wire fencing to surround the area to be sold.

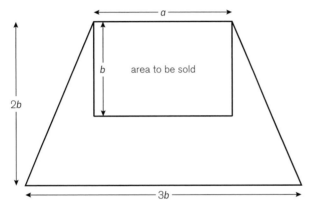

    Show that:
    a) $a + b = 100$
    b) $3b^2 + ab = 7500$

    Using these equations calculate the lengths a and b.

ACTIVITY 9

SOLVING MORE COMPLEX CONSTRUCTION PROBLEMS

In the real world many construction calculations require you to use a combination of methods. For example, to calculate the heat loss from a building will involve you calculating areas and volumes, but may also involve trigonometry and calculating arc lengths depending on the shape of the roof. So these examples use an combination of formulae and methods.

Worked example

Calculate the surface area of the roof. The building is 20 m long and a section through a building is shown here.

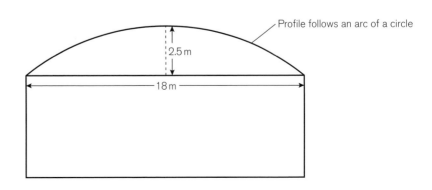

Profile follows an arc of a circle

2.5 m

18 m

1. Method

   The surface area of the roof will be the 'arc length' multiplied by the length (20 m).

   To calculate the arc length we need the radius of the arc and the subtended angle.

   To calculate the radius

   The first step is to sketch the radius onto the diagram.

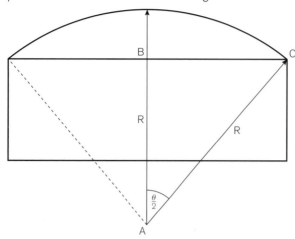

   We now have a right-angled triangle *ABC*. Its lengths are:

   $AC = R$ (radius)
   $BC = 9$ m (half the width of the building)
   $AB = R - 2.5$ m

   Given that we have the three sides of the triangle it should be possible to calculate the radius using Pythagoras.

   To calculate the angle θ

   We can use the same triangle *ABC* to calculate half the angle, which would then be doubled to obtain θ.

   Since all three sides are known, any of the sine, cosine or tangent ratios could be used to calculate the angle *BAC*. In this case, sine will be used.

2. Formulae

   Pythagoras is:

   $c^2 = a^2 + b^2$ ie $c = \sqrt{(a^2 + b^2)}$

   $\sin A = \dfrac{\text{opposite}}{\text{hypotenuse}}$

   Arc length $= R\theta \times \dfrac{\pi}{180}$

   Surface area = arc length × building length

3. Calculations

   To calculate the radius

   Substituting *AC*, *BC*, and *AB* into Pythagoras we have:

   $AC^2 = BC^2 + AB^2$

   Therefore:

   $R^2 = 9^2 + (R - 2.5)^2$
   $R^2 = 81 + (R - 2.5)\,(R - 2.5)$ [remember that $a^2 = a \times a$]
   $R^2 = 81 + R(R - 2.5) - 2.5(R - 2.5)$
   $R^2 = 81 + R^2 - 2.5R - 2.5R + 2.5^2$

   Collecting up the terms

   $R^2 - R^2 + 2.5R + 2.5R = 81 + 2.5^2$
   $5R = 81 + 6.25$
   $R = \frac{87.25}{5}$
   $R = 17.450$ m

To calculate the angle θ

$$\sin\left(\frac{\theta}{2}\right) = \frac{BC}{AC}$$

$$\sin\left(\frac{\theta}{2}\right) = \frac{9}{17.450}$$

$$\frac{\theta}{2} = \sin^{-1}\left(\frac{9}{17.450}\right)$$

$$\theta = 2 \times \sin^{-1}\left(\frac{9}{17.450}\right)$$

$$= 62.09644° \qquad [62° \ 05' \ 47.2'']$$

To calculate the arc length

$$\text{Arc length} = 17.450 \times 62.09644 \times \frac{\pi}{180} = 18.91209\,\text{m}$$

The arc length is **18.912 m**.

Note that the decimal value for the angle is used in the formulae and not its value degrees, minutes and seconds.

To calculate the roof surface area

$$\text{Area} = 18.91209 \times 20$$
$$= 378.2418\,\text{m}^2$$

The surface area of the roof is **378.242 m²**

4. Check
    A rough check would be to use the formula for a catenary mentioned in Activity 8. This is:

$$L = \frac{8S^2}{3d} + d$$

Where L is the arc length, d is the width of the building and S is the height of the roof.

$$L = \left(\frac{8 \times 2.5^2}{3 \times 18}\right) + 18$$

$$L = 18.92593$$

$$\text{Surface area} = 18.92593 \times 20$$
$$= 378.51852$$
$$= \textbf{378.519 m}^2 \ [\text{compared with } 378.242\,\text{m}^2]$$

This rough check would suggest that the value of 378.242 m² is correct.

Tasks

1. A building (6 m by 4 m) is to be set out relative to the fence line as shown in the diagram. The angles CDA and ABC are right angles. $AD = 2$ m; $BC = 4$ m; $CD = 6$ m.

    Calculate the lengths AB, AC and BD.

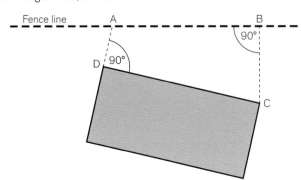

As a check try drawing it at a scale of 1:100.

2. During a survey of a house, the surveyor needs to calculate the height of the chimney. Using the measurements given below, calculate the height of the chimney from point B. Check your result from point A.

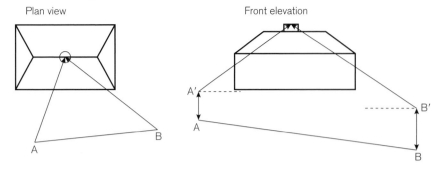

Plan view   Front elevation

Plan view	Front elevation
Horizontal angle at A is 64° 33′ 25″	Vertical angle at A' is 35° 22′ 43″
Horizontal angle at B is 43° 42′ 18″	Vertical angle at B' is 33° 48′ 50″
Slope distance AB is 9.645 m	Height AA' is 1.450 m
Point A is 1.250 m higher than point B	Height BB' is 1.550 m

ACTIVITY 10

USE STANDARD DEVIATION TECHNIQUES TO COMPARE THE QUALITY OF MANUFACTURED PRODUCTS USED IN THE CONSTRUCTION INDUSTRY

The 'average' does not give an indication of the quality of the data (see Activity 7). The standard deviation represents the spread of data from the arithmetic mean. The standard deviation is referred to by the symbol σ (which is pronounced sigma). The standard deviation is often used when manufacturing or measuring the quality and consistency of products. It might be used to assess the strength of different samples of concrete – the smaller the standard deviation, the greater the consistency of the concrete.

1. Standard deviation (σ)

Consider a batch of steel girders manufactured to 3 m ± 1.3 mm. The mean length of the girders should be 3 m and the '±1.3 mm' represents the tolerance – any girders outside of this tolerance would be rejected. The standard deviation (σ) is based on measuring a sample of actual girders.

At one standard deviation, we would expect 66 per cent of the girders to fall within the limits defined by the mean ± one standard deviation. To determine the limits for 95 per cent of girders, we would calculate the mean ±2σ. The limits for 99 per cent of the girders would be the mean ±3σ. These percentages are sometimes known as the confidence level.

If the standard deviation was 1.3 mm then we are fairly confident that 66 per cent of all the girders would be in the range 2998.7 mm to 3001.3 mm.

The formula for calculating the standard deviation of a set of data is:

$$\sigma = \sqrt{\frac{\Sigma(x - \bar{x})^2}{n}}$$

Where n is the number of items.

In construction, the standard deviation is usually based on a sample. For example, it would be impossible to test the strength of all the concrete used on a building. In practice, samples of the concrete are tested for strength. Therefore, it is important that you use this equation:

$$\text{Standard deviation (sample)} = \sqrt{\frac{\Sigma(x - \bar{x})^2}{n}}$$

Most scientific calculators will have statistical functions that enable you to calculate the standard deviation.

2. Grouped data

Sometimes it is convenient to group the data into bands or class intervals. For example, this table shows the frequency with which pile diameters fall into each interval:

Size in mm	1550.0 to 1550.4	1550.5 to 1550.9	1551.0 to 1551.4
Frequency	12	18	4

In this case we would use this formula to calculate the standard deviation:

$$\sigma = \sqrt{\frac{\sum f(x - \bar{x})^2}{\sum f - 1}}$$

Where f is the frequency of items, x is the midpoint of each class interval, $\bar{x}$ the mean of the data set. In grouped data, the mean is calculated by using this formula:

$$\bar{x} = \frac{\sum fx}{\sum f}$$

Worked example

A company manufactures 3 m steel girders. To check the quality, a sample of 50 girders has been measured. How accurately are these girders manufactured?

Length (m)	2.980–2.989	2.990–2.999	3.000–3.009	3.010–3.019
Frequency	5	26	10	9

1. Method

   The first step is to draw a table and calculate the midpoints. Then use the formulae to calculate the mean and standard deviation. The 95 per cent confidence level would be double the standard deviation. This means that 95 per cent of the girders would actually measure within the range of 3 m $\pm$(2 $\times$ the standard deviation in metres).

2. Formulae

   The mean is $\bar{x} = \frac{\sum fx}{\sum f}$ where f is the frequency of items.

   The standard deviation (sample) $= \sqrt{\frac{\sum f(x - \bar{x})^2}{\sum f - 1}}$ [f is the frequency of items]

   The 95 per cent confidence level is double the standard deviation.

3. Calculations

Class interval	Midpoint (x)	Frequency	fx	$(x - \bar{x})^2$	$f(x - \bar{x})^2$
2.980–2.989	2.98450	5	14.92250	0.000 213 160	0.001 065 800
2.990–2.999	2.99450	26	77.85700	0.000 021 160	0.000 550 160
3.000–3.009	3.00450	10	30.04500	0.000 029 160	0.000 291 600
3.010–3.019	3.01450	9	27.13050	0.000 237 160	0.002 134 440
Sum		50	149.955	Sum	0.004 042 000

The mean is $\bar{x} = \frac{\sum fx}{\sum f} = \frac{149.955}{50} = 2.9991$

The arithmetic mean is **2.999 m**

The standard deviation (sample) $= \sqrt{\frac{0.004\,042\,000}{50 - 1}} = 0.009\,082\,389$

The standard deviation is **0.009 m**

The 95 per cent confidence level is 2σ, that is: $2 \times 0.009\,082\,389 = 0.018\,164\,779$ m

So, the 95 per cent confidence level is **±0.018 m**

Therefore the quality of the steel girders is **2.999 ±0.018 m** at a 95 per cent confidence level.

4. Check
    The range of 2.999 ±0.018 m is from 2.981 to 3.017 m. Looking at the table, the majority of girders fall within this range.

Tasks

1. Two batches of bricks have been tested for strength (N/mm^2). Calculate the arithmetic mean, the standard deviation and the 95 per cent confidence level for each batch. Which batch is better in terms of consistency?

Batch A	Batch B
11.47	11.40
11.45	11.38
11.46	11.48
11.46	11.49
11.47	11.47
11.47	11.47
11.47	11.51
11.46	11.49
11.55	11.55
11.54	11.56

2. A next-door neighbour has built a two-storey extension blocking out the natural light in a property. A fenestration surveyor, an expert in window design, has been called in to measure the lighting levels in the kitchen of the affected property since. The surveyor's results are shown below.

Lighting level (lux)	50–99	100–149	150–199	200–249	250–299	300–349	350–399
Number of days	2	18	20	15	3	1	1

Calculate the arithmetic mean, the standard deviation and the 95 per cent confidence level.

ACTIVITY 11

USING SPREADSHEETS

It is important that you can independently check your final results. It is very easy to make a mistake when using a calculator, especially if you have to make repetitive calculations. One suggestion would be to use a spreadsheet such as Microsoft Excel to solve and check your calculations to complex problems. Calculations can be carried out by entering formulae into the spreadsheet's cells. It is important that formulae always start with an = sign.

A typical formula could be =A1+A2+A3+A4 or =SUM(A1:A4).

The most common functions used are:

+	add
−	subtract
*	multiply
/	divide
^	raised to the power
SQRT( )	square root
SUM(from:to)	adds up the contents of a range of cells
PI()	Pi (π)

Worked example

In Activity 4 (see worked example, problem 1) a table was produced to calculate the number of bricks required to build a garage. The same table can be reproduced in Excel. This illustration shows the formulae that would have been entered into the cells in column G.

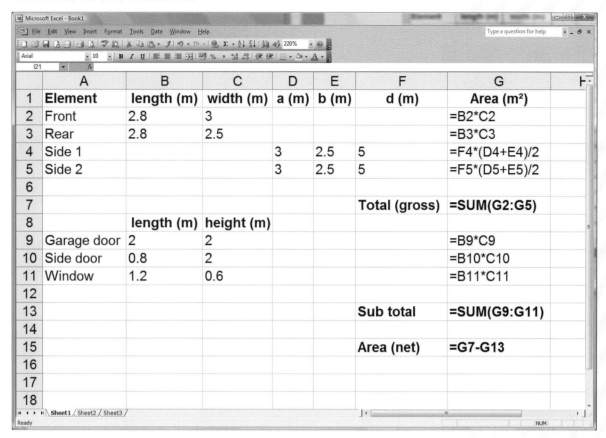

Microsoft product screenshot reprinted with permission from Microsoft Corporation.

And this is how the spreadsheet would normally look, showing the results.

115

	A	B	C	D	E	F	G	
1	**Element**	**length (m)**	**width (m)**	**a (m)**	**b (m)**	**d (m)**	**Area (m²)**	
2	Front	2.8	3				8.4	
3	Rear	2.8	2.5				7	
4	Side 1			3	2.5	5	13.75	
5	Side 2			3	2.5	5	13.75	
6								
7						**Total (gross)**	42.9	
8		**length (m)**	**height (m)**					
9	Garage door	2	2				4	
10	Side door	0.8	2				1.6	
11	Window	1.2	0.6				0.72	
12								
13						**Sub total**	6.32	
14								
15						**Area (net)**	**36.58**	
16								
17								
18								

Microsoft product screenshot reprinted with permission from Microsoft Corporation.

As you can see, it arrives at exactly the same results as we produced in Activity 4.

Tasks

1. Produce a spreadsheet that will calculate irregular areas using Simpson's rule and the trapezoidal rule (see Activity 4). Use these measurements.

Ordinate	Length
1	0.0
2	35.0
3	67.5
4	127.5
5	133.0
6	128.5
7	124.0
8	119.0
9	114.5
10	109.0
11	104.5
12	99.5
13	94.5
14	89.0
15	0.0

The distance between ordinates is 10 m.

Check your results with the worked example (problem 2) in Activity 4.

2. Produce a spreadsheet that automatically calculates the standard deviation (see Activity 10) using this data.

Length (m)	2.980–2.989	2.990–2.999	3.000–3.009	3.010–3.019
Frequency	5	26	10	9

Check your results with the worked example in Activity 10.

ACTIVITY 12

EVALUATING YOUR RESULTS

So far we have applied mathematical techniques to a variety of problems. You should now be able to apply these techniques in other modules, such as science, materials and surveying. However, before you present your results you need to ask yourself two key questions. First, have you made any calculation errors and, second, are the results realistic and appropriate?

Checking for calculation errors

In 1999 NASA lost a $125 million Mars orbiter because one engineering team used metric units and another used imperial units (feet and inches). This resulted in important information failing to transfer between the Mars spacecraft and mission control.

Two key points: first, always check the units and, second, check your calculations using a different method.

It is very easy to mix up the units in a situation where, say, room sizes may be in metres and window sizes in millimetres. Before you do any calculations, you must make sure all the measurements are in the same units.

Suppose you want to calculate the volume of a room 6.525 m by 8.105 m and 2500 mm high.

Volume = length × width × height

Volume = 6.525 × 8.105 × 2500 = 132,212.8125 m³   [this is wrong]

Volume = 6.525 × 8.105 × 2.500 = 132.2128125 m³   [this is right]

The calculations in both cases are correct and both answers contain the same digits, but see the difference the decimal point makes! A rough check calculation – such as 6.5 × 8 × 2.5 = 130 m³ – would confirm that the 132 m³ is the right answer.

A further issue is the number of decimal places quoted. If your measurements are taken to three decimal places then your final answers cannot be to a higher precision, ie four decimal places or more.

Checking that the results are realistic and appropriate

Consider the example in Activity 5. The sextant angle calculated was 14° 02′ 10″ but this is not a realistic answer because sextants cannot measure to 10″ of arc. A better solution would be to have a set angle of say 15° and vary the marker posts on shore. To calculate the angle at the end of the pier we have:

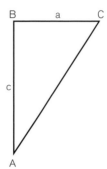

To calculate BC:

$$\tan A = \frac{BC}{AB}$$

$BC = AB \times \tan A$

$\quad = 200 \times \tan 15° = 53.590 \, m$

Angle A 15° fixed
Adjacent side (c) is 200 m
Opposite side (a) is unknown

Therefore, instead of setting out BC to 50 m, a better solution would be to set out BC to 53.590 m. This is a more practical solution to the problem. It is this type of questioning and analysis that you need to develop.

Tasks

1. Assessment of calculating irregular areas
    In Activity 4 we calculated the area of some allotment gardens based on an Ordnance Survey plan.

    a) How accurately can you measure distances on a 1:1250 plan?

    b) How would you check that the plan had not been distorted by poor photocopying?

    c) What alternative methods could be used to calculate the area more accurately?

2. Assessment of measuring angles
    In Activity 9 (task 2) you were asked to calculate the height of a chimney. A theodolite is an instrument that is used to measure vertical angles. In the two scenarios below, estimate how accurately the theodolite would need to measure the vertical angle to ensure that the height of the chimney and tower are accurate to ±0.010 m.

Building   Tower

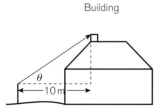
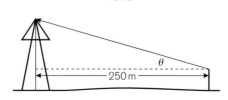

ANSWERS

ACTIVITY 1

1. 4.100

2. 5.033

3. −0.514

4. 0.431

5. 23.664

6. 494.173

ACTIVITY 2

1. $r = \sqrt{\dfrac{A}{\pi}}$

2. $b = \sqrt{(c^2 - a^2)}$

3. $y = \dfrac{x}{2}$

4. $k = \dfrac{dH}{tA(\theta_1 - \theta_2)}$

5. $\cos A = \dfrac{b^2 + c^2 - a^2}{2bc}$

6. $B = \sin^{-1}\left(\dfrac{b \sin A}{a}\right)$

ACTIVITY 3

Task 1

Temperature	Distance
−5	50.014
50	49.983

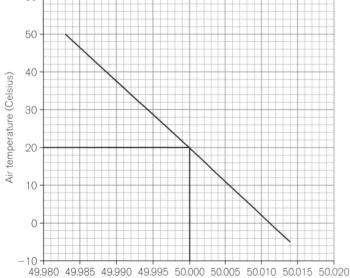

Tape temperature correction

118

Task 2

Mile	Unit	km
0.00	0.00	0.00
6.21	10.00	16.09
12.43	20.00	32.19
18.64	30.00	48.28
24.85	40.00	64.37
31.07	50.00	80.47
37.28	60.00	96.56
43.50	70.00	112.65
49.71	80.00	128.75
55.92	90.00	144.84
62.14	100.00	160.93

35 miles = 56.33 km
80 km = 49.71 miles

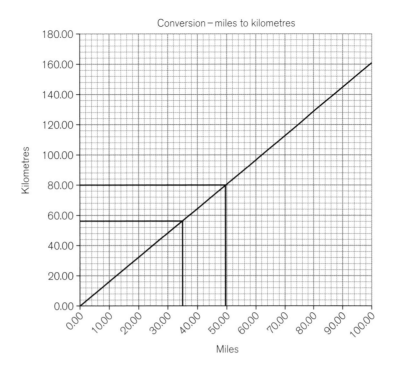

Task 3

h	y
3.00	180.465
3.05	170.947
3.10	161.296
3.15	151.511
3.20	141.593
3.25	131.543
3.30	121.362
3.35	111.050
3.40	100.609
3.45	90.038
3.50	79.340
3.55	68.514
3.60	57.561
3.65	46.482
3.70	35.278
3.75	23.949
3.80	12.497
3.85	0.921
3.90	−10.776
3.95	−22.595
4.00	−34.535

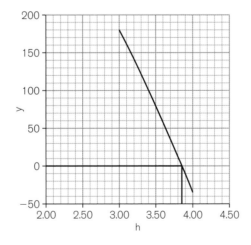

$h = 3.85\,m$

Task 4

h (m)	y
0.150	−1.890
0.160	−1.422
0.170	−0.956
0.180	−0.490
0.190	−0.024
0.200	0.440
0.210	0.904
0.220	1.366
0.230	1.828
0.240	2.290
0.250	2.750

$h = 0.19$ m

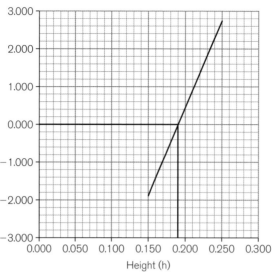

ACTIVITY 4

Task 1

	Area	
	45.030	[Area 1]
	100.080	[Area 2]
	135.000	[Area 3]
	104.550	[Area 4]
	371.175	[Area 5]
Total	755.835	m²
Volume	113.375	m³

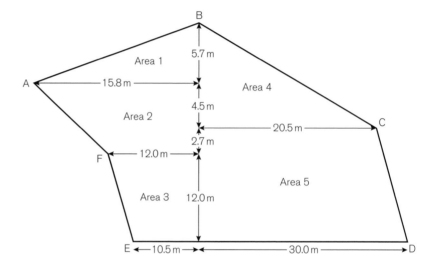

Task 2

a) Volume of roof space

$$\text{Volume} = \frac{h}{3}(A_{\text{base}} + A_{\text{top}} + \sqrt{(A_{\text{base}}A_{\text{top}})})$$

A_{base} is the area of the base $= 6 \times 4 = 24$
A_{top} is the area of the top $= 2 \times 4 = 8$
h is the perpendicular height $= 2.828$
$V = \frac{2.828}{3}(24 + 8 + \sqrt{24 \times 8}) = 43.22730583$
$V = \textbf{43.227 m}^3$

b) Surface area of roof

$$\text{Slanted surface area} = \frac{a(P_{\text{base}} + P_{\text{top}})}{2}$$

P_{base} is the perimeter of the base $= 4 + 6 + 4 + 6 = 20$
P_{top} is the perimeter of the top $= 2 + 4 + 2 + 4 = 12$
a is the slant length $= 3$
Slanted surface area $= 3(20 + 12)/2 = 48$
Area of top $= 2 \times 4 = 8$
Total area $= 8 + 48 = 56$ m²

Task 3

A	0.0	First
B	0.8	Even
C	1.0	Odd
D	1.2	Even
E	1.3	Odd
F	1.1	Even
G	1.1	Odd
H	1.0	Even
I	1.4	Odd
J	1.3	Even
K	1.3	Odd
L	1.2	Even
M	1.3	Odd
N	1.4	Even
O	1.5	Odd
P	1.6	Even
Q	0.0	Last

Sum first and last   = 0.0
Sum odd ordinates   = 8.9
Sum even ordinates   = 9.6
Distance between (d)   = 2

By Simpson's rule area   = 37.467 m^2
By trapezoidal rule area   = 37.000 m^2

Task 4

Section	Area	Comment
1	1485	First
2	736	Even
3	300	Odd
4	91	Even
5	15	Last
Sum of odd areas	300.0	
Sum of even areas	827.0	
Sum of 1st + last areas	1500.0	
Distance	2.0	
Simpson's	**3605 m^3**	
Trapezoidal	**3754 m^3**	
Difference	**−149 m^3**	

Sum first and last   = 1500.0
Sum odd ordinates   = 300.0
Sum even ordinates   = 827.0
Distance between (d)   = 2.0

By Simpson's rule volume = 3605 m^3
By trapezoidal rule volume = 3754 m^3
Difference = 149 m^3

ACTIVITY 5

Task 1

$CD = 4.5 \times \tan 30°$ $= 2.598\,m$

$AC = 4.5 \div \cos 30°$ $= 5.196\,m$

$AB = BC = 5.196 \div 2$ $= 2.598\,m$

$BE = 2.598 \times \tan 30°$ $= 1.500\,m$

$AE = CE = 2.598 \div \cos 30° = 3.000\,m$

Task 2

Height of tree $= (20 \times \sin 19.5°) + (20 \times \cos 19.5° \times \tan 35°)$

Height of tree $= 19.877 = 19.9\,m$

The distance from the tree to the garage $= (20 \times \cos 19.5°)= 18.85\,m$. So if the tree falls, its top will hit the garage.

ACTIVITY 6

Task 1

The kerb is a sector of a circle with angle θ and radius R. Using simple geometry, it can be shown that $\theta = 125°$. Using trigonometry, we have:

Chord length $= 2R \sin\left(\dfrac{\theta}{2}\right)$

$R = \dfrac{\text{chord length}}{2 \sin\left(\dfrac{\theta}{2}\right)} = \dfrac{5.550}{2 \sin\left(\dfrac{125}{2}\right)}$

Radius $= 3.128\,m$

Arc length $= R\theta° \times \left(\dfrac{\pi}{180}\right) = 3.12848 \times 125 \times \dfrac{\pi}{180}$

Arc length $= 6.825295 = 6.825\,m$

Task 2

a) The cross-sectional area $= 7773.3\,mm^2$

b) The wetted perimeter $= 235.6\,mm$

c) The hydraulic mean radius $= 33.0\,mm$

ACTIVITY 7

Task 1

	Quantity surveyors	Bricklayers
Median	£2,594.54	£1,635.01
Arithmetic mean	£2,725.81	£1,680.91
Geometric mean	£2,696.37	£1,661.11

When dealing with earnings, the median is probably the best method because it excludes the highest and lowest values thus making the average more realistic.

Task 2

Bar chart

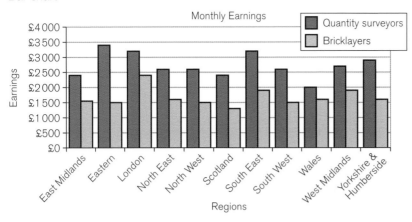

Line diagram

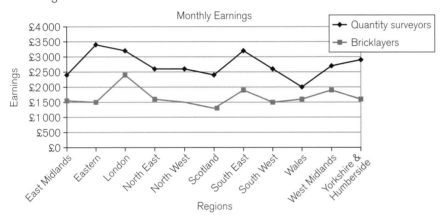

Some observations

- Bricklayers earn significantly more in London than the rest of the UK.
- London bricklayers even earn more than quantity surveyors in three regions.
- The difference in earnings between bricklayers and surveyors is greatest in the Eastern region and smallest in Wales.

Task 3

a) Missing out or compressing the years (*x*-axis) suggests a much steeper rise than what actually occurred. This would be a better graph.

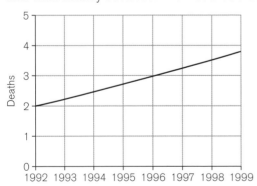

b) • The sales axis (*y*-axis) has no unit values. It could be the number of items sold or the value (£).
- The *y*-axis has unequal values.
- The *x*-axis has missing months.
- The bar for July is shaded but we don't know if this is significant or not.

c) • No scale to say how many sites are represented by each symbol.
- Do not know the significance of the size of each symbol.

ACTIVITY 8

Task 1

Convert the equation into a quadratic:

$$3d^2 - 180d + 24.5 = 0$$
$$\text{So, } a = 3, b = -180, c = 24.5$$

Using the formula to solve a quadratic, $d = 0.136\,\text{m}$ (unrealistic) or $59.864\,\text{m}$ (correct answer).

Task 2

Perimeter of area to be sold $= 2a + 2b$
So $2a + 2b = 200$, hence $a + b = 100$

Area of trapezium $= \dfrac{(a + 3b)}{2} \times 2b = 7500$

So $(a + 3b)\, b = 7500$, hence $3b^2 + ab = 7500$

Combine the equations to obtain a quadratic:

$3b^2 + (100 - b)\, b = 7500$, hence $2b^2 + 100b - 7500 = 0$

Use the quadratic formula to obtain $b = 41.144\,\text{m}$ and $a = 58.856\,\text{m}$

ACTIVITY 9

Task 1

$AC = \sqrt{2^2 + 6^2} = 6.325\,\text{m}$

$AB = \sqrt{AC^2 - 4^2} = 4.899\,\text{m}$

Angle $C = \tan^{-1}\left(\dfrac{AD}{DC}\right) + \cos^{-1}\left(\dfrac{BC}{AC}\right) = 69°\ 12'\ 12''$

$BD = \sqrt{DC^2 + BC^2 - 2 \times DC \times BC \times \cos C} = 5.912\,\text{m}$

Task 2

Horizontal distance $AB = R9.645^2 - 1.250^2 = 9.563\,656\,\text{m}$

Angle at chimney $(C) = 180° - (64°\ 33'\ 25'' + 43°\ 42'\ 18'') = 71°\ 44'\ 17''$

Horizontal distance $BC = \dfrac{\text{horizontal distance } AB}{\sin(\text{horizontal angle at } C)} \times \sin(\text{horizontal angle at } A) = 9.094\,140\,\text{m}$

Height of C from B = (horizontal distance $BC \times \tan$(vertical angle at B')) + 1.55

Height of C from B = $(9.09414 \times \tan (33°\ 48'\ 50'')) + 1.55 = 7.641\,\text{m}$

So the height of C from A should be 6.391 m (as we are told that A is 1.250 m higher than B). To check this by another method, calculate the horizontal distance AC.

Horizontal distance $AC = \dfrac{\text{horizontal distance } AB}{\sin(\text{horizontal angle at } C)} \times \sin(\text{horizontal angle at } B) = 6.958\,435\,\text{m}$

Height of C from A = (horizontal distance $AC \times \tan$(vertical angle at A')) + 1.45

Height of C from A = $(6.958\,435 \times \tan (35°\ 22'\ 43'')) + 1.45 = 6.391\,\text{m}$

ACTIVITY 10

Task 1

	Batch A	Batch B
Arithmetic mean	11.480	11.480
Standard deviation	0.035	0.057
95% confidence level	0.070	0.114

Note that the mean for both batches is the same. Batch A is better in terms of consistency because the standard deviation is smaller.

Task 2

Class interval	Midpoint (x)	Frequency (f)	fx	$(x - mean)^2$	$f(x - mean)^2$
50–99	74.50	2	149.00	11 025.00	22 050.00
100–149	124.50	18	2241.00	3025.00	54 450.00
150–150	174.50	20	3490.00	25.00	500.00
200–249	224.50	15	3367.50	2025.00	30 375.00
250–299	274.50	3	823.50	9025.00	27 075.00
300–349	324.50	1	324.50	21 025.00	21 025.00
350–399	374.50	1	374.50	38 025.00	38 025.00
	Sum	60	10 770.00		193 500.00
	Mean	179.500			
	Std. dev. (σ)	57.268	67%		
	2σ	114.537	95%		

The results show that lighting levels were 180±115 lux (ie from 65 to 295 lux) for 95 per cent of the samples.

ACTIVITY 11

Task 1 – suggested layout

Formula screen

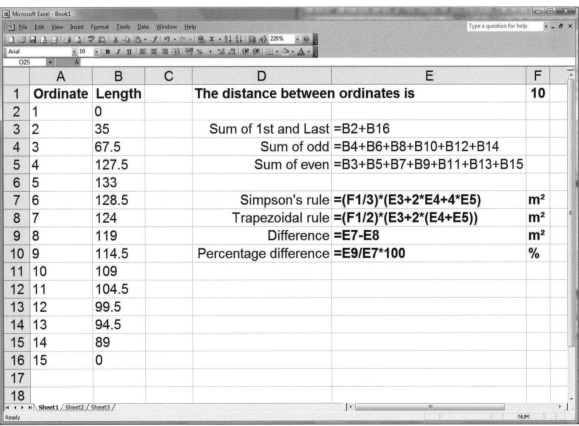

Microsoft product screenshot reprinted with permission from Microsoft Corporation.

Results screen

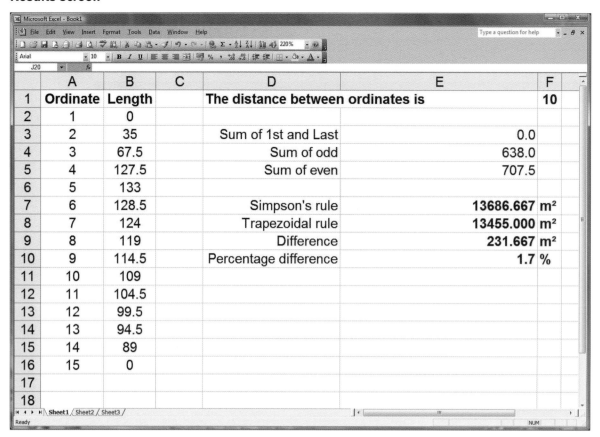

Microsoft product screenshot reprinted with permission from Microsoft Corporation.

Task 2

Formula screen – suggested layout

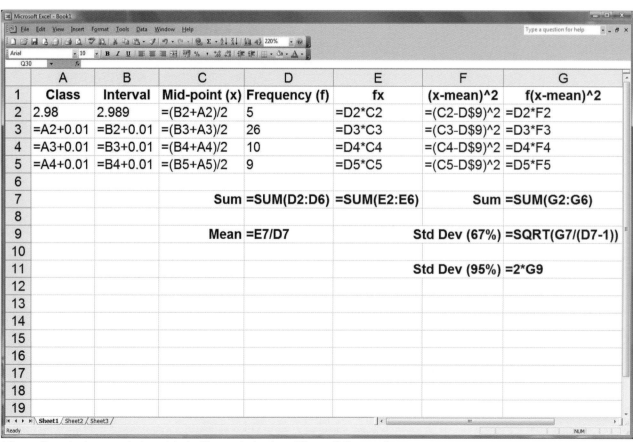

Microsoft product screenshot reprinted with permission from Microsoft Corporation.

Results screen

	A	B	C	D	E	F	G
1	**Class**	**Interval**	**Mid-point (x)**	**Frequency (f)**	**fx**	**(x-mean)^2**	**f(x-mean)^2**
2	2.980	2.989	2.9845	5	14.9225	0.00021316	0.00106580
3	2.990	2.999	2.9945	26	77.8570	0.00002116	0.00055016
4	3.00	3.009	3.0045	10	30.0450	0.00002916	0.00029160
5	3.010	3.019	3.0145	9	27.1305	0.00023716	0.00213444
6							
7			**Sum**	**50**	**149.9550**	**Sum**	**0.00404200**
8							
9			**Mean**	**2.9991**		**Std Dev (67%)**	**0.009**
10							
11						**2 x Std Dev (95%)**	**0.018**

Microsoft product screenshot reprinted with permission from Microsoft Corporation.

ACTIVITY 12

Task 1

a) Realistically you can scale to 0.5 mm which at 1:1250 represents 625 mm or 0.625 m.

b) Every map should have a grid or scale bar, so you would check the error. You would adjust your measurements proportionally.

c) Use an instrument called a planimeter. The most accurate method would be to overlay the area with graph paper and count the millimetre squares.

Task 2

The building

The minimum angular error would be $\tan^{-1}\left(\dfrac{0.01}{10}\right) = 0° \, 03' \, 26''$

Therefore angles should be measured better than $\pm 3'$ of arc.

The tower

The minimum angular error would be $\tan^{-1}\left(\dfrac{0.01}{250}\right) = 0° \, 00' \, 08''$

Therefore angles should be measured better than $\pm 8''$ of arc.

UNIT 7 – PLANNING, ORGANISATION AND CONTROL OF RESOURCE IN CONSTRUCTION AND THE BUILT ENVIRONMENT

This section focuses on the factors that contribute towards managing a successful construction or civil engineering project. It looks at the resources and techniques that are used to turn the designer's drawings and specifications into reality, ensuring that the project is achieved on time, on budget and to the appropriate quality. This section focuses on P1, P2, P3 and M3 and contributes to D2

In this unit you will learn about the importance of planning, organising and controlling resources in the production of construction and civil engineering projects. The factors that affect the quality, cost and timely completion of a project will be explored and you will learn some basic techniques and methods that lead to successful construction projects.

Content

1) **Understand the roles and responsibilities of, and interaction between, the parties involved at each stage of the construction process**

   Roles and responsibilities: management at director or site level; technical roles, eg as planner, quantity surveyor, buyer, estimator; supervisory roles, eg contract supervisor, general foreperson, general operative roles; craft roles, eg joiner, bricklayer, steel fixer.

   Team working and interaction of roles: head office and site organisational charts; team interaction and communication (communication methods, instruction, co-operation, co-ordination, control); levels of responsibility and accountability.

   Stages of construction: elements of a typical low-rise domestic or commercial building; design; production; maintenance; alteration; refurbishment; demolition.

   Planning the construction process: feasibility studies; design; procurement; production; maintenance and repair; refurbishment.

2) **Be able to identify the resources required to complete a construction project and describe how each is used**

   Resources: human (direct and subcontract labour, management and supervision); plant and machinery; materials; subcontractors.

   Factors in the planning process: labour factors, eg availability and cost, skill levels, motivation, productivity; plant factors, eg output rates and efficiency; material factors, eg availability, delivery periods, site handling, waste.

   Context: finance; site layout and organisation; temporary facilities and works; health, safety and welfare issues.

   Uses: production of long-term and short-term programmes; scheduling of material requirements; requisitioning and ordering; receiving and checking; site handling; storage and security issues; labour management techniques (work and method study, control and organisation of labour); plant management (hire, lease or purchase, utilisation and control); relevant documentation.

3) Understand the functions of management in the organisation of the production stage of a construction project

General functions: forecasting; planning; organising; monitoring; controlling; co-ordinating; reviewing.

Organisational aspects: site layout plan; traffic routes; labour movement; materials and plant location; access and egress; site accommodation; storage; security; health, safety and welfare; method statements; progress monitoring; site meetings; subcontractor liaison; site resources; documentation control; programmes of work; bar charts; schedules, eg line of balance; network diagrams, eg arrow diagrams.

Variables: weather; availability of skilled labour; labour disputes; confined access; late design changes; late construction information; material shortages.

Grading criteria

P1 *identify and describe the various stages of the construction process for a low-rise domestic or commercial building*

When planning, designing and constructing a building it is important to break down the project into clearly separate stages so that it can be built on time, on budget and to the right quality of materials and workmanship.

To meet this criterion you must identify all stages from design, through construction to handover of a completed building. For each stage you will need to show its key features; for the construction stages ensure that you include all the main construction elements together with all the temporary work that is required.

P2 *investigate and explain the roles and inter-relationships of those members of the building team involved in resource management, planning and production*

The process of construction relies greatly on good communication and management skills being used in resource management, planning and production. To meet this criterion you must identify and explain the main roles within the design and construction teams and how they interact with each other. The key aspects of this interaction should be identified and described including communication, instruction, co-operation, co-ordination and control.

P3 *identify the physical and human resources required to complete a building project*

It is important that you have knowledge and understanding about the craft skills, labour, plant, materials and equipment that are used to construct a building. To meet this criterion you will need to demonstrate that you can recognise these resources and the important part they play in the carrying out of building works.

M3 *discuss the factors that may adversely impact on planning and organisation if not taken into consideration, and explain their possible effects*

When planning and organising any construction activity it is important to try and foresee any problems so that they can be avoided or properly managed. This criterion requires you to discuss possible problems and their effects on a construction project. Examples include poor weather, changes to the design, changes to the construction method or materials, and factors to do with supply and delivery.

D2 *evaluate a range of planning, organisational and control techniques in terms of utility and efficacy*

Good site managers very closely monitor the progress of work on their construction sites using a variety of systems and documentation. This criterion involves you examining some of these planning, organisational and control techniques and exploring how useful and efficient they are in given situations.

129

ACTIVITY 1

THE CONSTRUCTION TEAM AND THEIR ROLES

This activity will involve you exploring the people who make up the construction team. You will create an interactive poster that will help you learn what their responsibilities are, the functional operations that they carry out and how different levels in the team interact.

Task 1

Consider this list of job titles of people who are part of the construction team:

- director or chief executive
- contracts manager
- site manager
- estimator
- buyer
- planner
- cost control surveyor
- health and safety consultant/officer
- document controller
- subcontractor (manager)
- general foreperson (man/woman)
- craft operative (carpenter, plumber, etc)
- manual operative (groundworker, gatekeeper etc)

Using the ConstructionSkills website at www.bconstructive.co.uk and other sources such as textbooks, make rough notes on the individual roles and responsibilities of each job listed.

Task 2

A construction team's work can be broken down into three different but important operations.

- **Construction planning**
  Examining the logical sequence and timing of the work, and reviewing the best way of carrying out the work safely and efficiently.

- **Resource management**
  Examining the resources needed to construct the work; quantifying, locating, buying, hiring, ordering and also administering the finances for them.

- **Construction production**
  Safely carrying out the physical processes of construction using plant, materials, and skilled and unskilled labour.

Obtain some white card and cut it into rectangles of approximately half the size of a postcard. Obtain three coloured felt tip pens: one red, one orange and one green.

Produce an information card for each job title listed in Task 1. Print the job title neatly in block capitals on the card. Then add the definition of that job's roles and responsibilities, which you researched in Task 1. Finally, match the job roles against the three operations of a construction team by using your felt tip pens to tick ✓ each card as follows:

- all those that relate to **construction planning** tick with a green coloured pen
- all those that relate to **resource management** tick with an orange coloured pen
- all those that relate to **construction production** tick with a red coloured pen.

If you feel that a job title involves a combination of two operations, tick the card with both the appropriate coloured pens. If you feel that one role involves all three, then tick the card accordingly with three colours!

Task 3

Study your cards from Task 2 and sort them into one of these five groups.

Designation	Description
Professional	Chartered member of a professional body such as the Chartered Institute of Building (CIOB) with management powers over staff and resources, and some responsibilities for the development of a construction company.
Technical	Incorporated member of a professional body with technical skills, knowledge and expertise; may oversee an element of the construction of the project and be in charge of a small team of people.
Supervisory	Trained and experienced member of the team who is in charge of practical work; monitors work, sources materials and assists the craft operatives to achieve the quality required within the time and budget set out in the contract.
Craft	Qualified and trained in a practical skill (to NVQ level 2 or 3) and also in possession of suitable 'time-served' experience.
Labourer	Supplies manual labour; relatively unskilled.

Get a large sheet of paper (at least A2 size), and list these five designations on the sheet. List them in order, the top tier being 'Professional', the tier below 'Technical', and the tier below that 'Supervisory', etc. Stick the shaded cards (from Task 2) in the correct tier on your A2 sheet with glue. Title up the A2 sheet with the heading 'Roles and responsibilities in the construction team', then hang it on the wall.

Task 4

Problems for Bestend Contracts

On a site close to the town centre, a main construction contractor Bestend Contracts is building a major mixed development of shops and offices. Its client is a major financial investment company called Richlitz. The site itself was once a railway goods yard and shunting area. It is the third week in November and there has been a problem encountered during the excavation of the new foundations to Block A. Some old train shed foundations have unexpectedly been discovered by Dufflin, the groundworks subcontractor. These old obstructions are very deep, and they cannot be removed by the excavation plant that is currently available on site, and there is the possibility of further obstructions being found.

Possible solutions are:

- to bring in larger excavation plant to try to remove the obstructing foundation
- to redesign the new foundations around the obstructions
- to redesign and reposition Block A on the site to avoid the obstructions completely.

Bestend Contracts is currently on programme to complete the works on time but this unexpected delay is going to cause problems.

The architect Marcus Hutching, who works for the Rackman Design Partnership, is chairing his weekly site meeting to review this problem. Also in attendance at the meeting are:

Jack Barrymore	Bestend's contracts manager
Steve Jones	Bestend's site manager
Rebecca Donovan	Bestend's planner and cost control surveyor
Paul Hudson	Principal of Dufflin, groundworks and foundation subcontractor
Alf Richards	Groundwork foreman for Dufflin
Hassam Chowdary	Consultant structural/civil engineer working for WRP Consultants

Read the scenario carefully and then answer these questions:

a) If this is a traditional project as outlined in the RIBA Plan of Work (see page 134, Task 3), what is the role of Marcus Hutching?

b) Describe the organisational relationship between Bestend's staff at this meeting.

c) What matters do you think Alf Richards would report on at this meeting? And how does his contribution relate to Paul Hudson's role?

d) What would have been the likely involvement of WRP in this project to date?

e) Who is responsible for the overall programme of works and for ensuring that all resources are in place to build the development? Who would evaluate the alternative solutions in terms of their impact on the overall construction programme?

f) What factors will govern the choice of solution to this unexpected problem? And who would investigate these factors?

g) Explain who would have the most input in deciding on which solution to use.

Task 5

In writing, define the following terms that are used in managing people and situations. They are sometimes called 'management tools'. Check your definitions using a dictionary.

- Communication
- Instruction
- Co-operation
- Co-ordination
- Control

Review your responses to Task 4 above, and for each of the people described in the scenario identify and describe how they relate to these five management tools.

ACTIVITY 2

WHAT IS A CONSTRUCTION PROJECT?

Task 1

The two lists shown below – 'Title' and 'Descriptions of construction phases' – have been jumbled up.

a) Photocopy the lists onto a piece of thin card then cut up each individual 'title' and 'description' of the construction phases into separate rectangles or tiles.

b) Working in pairs, arrange the rectangles on a flat surface so that the correct title goes with the correct description. Make sure that you both agree that a pairing is correct, before you go on to the next card.

Title	Description of construction phase
Construction operations	A. At the end of the construction works the building designer will carry out a survey to ensure that the work is complete and constructed to the specified quality. If there are any problems or snags, then these have to be corrected before the completion certificate can be issued and the contractor is paid the final instalment which usually includes his profit from the contract. The quantity surveyor may adjust the final sum due to the contractor depending on whether there had been any variations by the client or considerations under the conditions of contract, such as compensating for unavoidable bad weather.
Final scheme	B. The client, building designer and quantity surveyor review the tenders and award the building contract to the main contractor that they believe offers the best value in terms price, cost and quality of work.
Decommissioning	C. This is where the bulk of the physical work is done. The main contractor is responsible for all site operations including health and safety and insurance. The contractor must ensure that the work keeps to the programme and the budget for the project. The quality of the work must match the original specification and this is often checked by a clerk of works or resident engineer, but the contractor does not want to do abortive work so it is important that the output of the workforce (including all subcontractors) is constantly inspected. The contractor will be paid in stages throughout this phase to provide a cash flow to finance the later stages of the works.
Client's brief	D. The building designer uses the brief to develop the design(s) of the project in the form of labelled sketches and a report. This may require further research and investigation with the help of other specialists within the design team, such as the structural engineer and the quantity surveyor. The client approves the scheme.
Brownfield site redevelopment	E. The client appoints, then works with, a building designer to formulate a document called the brief. This sets out the client's requirements, any possible legal or financial constraints, and a time scale for the project.
Plan and mobilise for construction	F. Once a building has been demolished, the land can be reused to construct the new buildings. This process can regenerate an area that was previously run down. Care has to be taken to ensure that there are no harmful residues, such as asbestos or oil, left in old industrial or commercial buildings. A client may buy this type of site in order to acquire land to put up a new building.
Maintenance	G. The design team works together to refine detailed designs and also to seek approval from the local authority planning and building control departments. The quantity surveyor continues to advise the client of the likely costs of the project.
Awarding the contract	H. The final drawings, bills of quantities and written specification for the project are sent out to selected main contractors. These contractors will study the details and estimate how much the work will cost to do and how long it will take to build. They have to allow for the cost of materials, plant and labour as well as their overheads and, of course, their profit.
Sketch designs	J. The client takes possession of the building and receives a building manual which includes 'as-built' drawings, service layouts and emergency control procedures. It also includes important information about how the building can be safely cleaned, maintained, extended and demolished. In a domestic home, the maintenance work is usually undertaken by the owner; in large office buildings, schools and hospitals it is usually organised and undertaken by the facilities manager and a team of workers.
Refurbishment and renewal	K. The main contractor assembles the construction team, including subcontractors and consultants. Theteam carries out pre-construction activities, which include the detailed programme planning for the works, negotiates with suppliers and starts placing orders for materials and plant. It sets up secure site offices and organises the necessary temporary services, such as telecommunications, power and water.
Tendering for a contractor	L. When a building no longer meets its original function, it can sometimes be altered to meet a new use. However, if it is uneconomic to repair or refurbish it, it may be demolished. The old materials can often be reused or recycled in new buildings, and specialist contractors are used to manage these operations. Some historic buildings are 'listed' – they cannot be demolished by law, so other solutions have to be found to preserve them. In some cases, this can even involve moving them to other locations.
Defects period and final completion	M. The building is designed for a specific life span, depending on the materials used in construction and the expected 'wear and tear', due to the use of the building by its occupants. To extend the life of the building, its elements will need to be inspected at regular intervals and repaired. It may also be necessary to renew services such as the electrical wiring (particularly if regulations change).

Task 2

a) Working in your pairs, study the groupings of tiles that you have created in Task 1. Which pair do you think comes first in the sequence of the life cycle of a construction project? Do you both agree?

b) Then starting from this pair, and keeping the rest of the tiles in their pairs, lay the cards out in a correct logical sequence. There should be 12 pairs, so lay them out like a clock face, proceeding clockwise from start to finish. Follow the process through and make sure that you both agree on the logical sequence for all 12 phases. Be sure that you are both happy with your final layout.

c) Take a photograph, using a digital camera or other device, to record the 'life cycle clock'. This should show the life cycle of a construction project. Regardless of your starting point on the clock face, this construction process will repeat itself.

Task 3

The Royal Institute of British Architects (RIBA) publishes a useful guide to show how the design of buildings relates to the construction processes. This guide is known as the RIBA Plan of Work. It takes the life cycle that we have studied in the Tasks 1 and 2 and classifies the work into five sequential stages.

1. Feasibility stage
2. Design stages
3. Pre-construction stages
4. Construction stages
5. Review stage

Undertake research to write a brief definition of these stages. A useful resource to use here is the RIBA website (www.architecture.com), which is also where you can download a copy of the RIBA Plan of Work. Alternatively you could use any good construction technology textbook or glossary.

Task 4

In this unit, you will be concentrating on the work of the construction team during the pre-construction, construction and review stages of the process. The feasibility and design stages are largely the work of the architect and members of the design team, which is covered in Unit 5 of this course.

Stewards Elm Lane de luxe housing development

A developer wishes to build seven large detached houses on a large site on the edge of a small town (see site plan below). Each house will be two storey and will have its own integral double garage.

The site is poorly drained and largely flat. The existing trees are to remain and must be protected throughout the construction works. A footpath runs on the northern boundary next to the stream, the southern boundary is a narrow public lane. The main services including the sewers are also accessed from the lane. Electric and telephone cables run under the foot pavement.

The houses are to be constructed of cavity brickwork and blockwork walls, precast suspended pot and beam ground floors, timber first floors and trussed timber roofs. The foundations will be concrete strip foundations throughout, except for those built next to the oak tree which will have piled foundations.

Here is a list of the pre-construction activities that the contractor needs to carry out for this project prior to construction. They are in no particular order.

- Bring plant and equipment to site
- Erect hoardings or fences
- Lay on temporary electric, communications, water and drainage services
- Deliver and install site offices
- Set out the foundations for new buildings, and erect profiles
- Deliver and install welfare facilities such as toilets and canteen
- Determine ordering dates for prefabricated components
- Plan out the detailed construction programme
- Establish the required delivery dates for materials
- Carry out site survey to confirm shape, levels and position of boundaries
- Sign contracts and take possession of the site
- Place order for specialist subcontractors
- Install temporary roads and pedestrian access
- Clear site of previous building(s) and rubbish

Make sure that you understand what each of the preconstruction activities involves. Then rearrange them by drawing a flow chart to indicate the order in which they could be carried out – you should start with 'Sign contracts and take possession of the site'. Note that some activities can happen at the same time; in other words, there could be parallel activities.

Task 5

It is important to be familiar with the terms and technology associated with the type of construction proposed at Stewards Elm Lane (see Task 4). This task will enable you to access an online learning activity that takes you through the basic construction used in floor, wall and roof construction methods.

The activity can be found on the National Learning Network website. All recognised sixth forms and further education colleges are members of the National Learning Network and will have access to these online activities. Your school or college may already have these materials linked to its 'virtual learning intranet' – if not, ask your tutor for the organisational password; you will need this to log into the site.

To access the activity, go to the National Learning Network home page (www.nln. ac.uk). Then click through the tabs 'Vocational' then 'Construction and Engineering L3' and then browse through the list of activities for the one entitled 'Primary elements of a building'. The activities are in alphabetical order.

Alternatively you could carefully type this URL into your address bar of your web browser:

http://go.nln.ac.uk/?req=%7B3F9AAE48-5982-43FF-BAA8-3C489D6F6DDE %7D&prov=aclProv2264

Take time to read the instructions and then work through the interactive routines and quizzes based around the 'ground floor', 'upper floor', 'walls' and 'roof' activities.

Task 6

It is important to understand the main construction operations that form the basis of a contractor's planning and work

Some examples of construction operations are given below, listed under four primary headings:

A. Construct the substructure
- excavate and concrete the foundations
- lay the foul and storm drainage

B. Construct the superstructure
- build the ground floor
- construct the internal and external walls
- fix the upper floors
- construct the roof

C. Install the building services
This is usually done in two stages, for example:
- first fix – fixing all the cable and pipes runs etc
- second fix – fixing all the service fittings such as plug sockets, light fittings and control systems

D. Construct the external works and landscaping
- lay external driveways and patios
- clear the site of contractor's facilities, and landscape garden

This table shows a break down of the construction operations required by the Stewards Elm Lane de luxe housing development:

Operation reference	Construction operations	To be preceded by operation
1	Site set up	–
2	Strip site and setting out building	1
3	Excavate foundations	2
4	Concrete foundations	3
5	Excavate drainage	3
6	Lay and bed drainage	5
7	Backfill drainage trenches	6
8	External walls up to dpc and ground floor	4
9	Construct concrete ground floor	6 and 8
10	External walls up to first floor level	9
11	Internal partitions at ground floor	9
12	Construct first floor timbers	10 and 11
13	Lay first flooring chipboard	12
14	External walls up to roof eaves level	10

continued on next page

Operation reference	Construction operations	To be preceded by operation
15	Internal partitions at first floor	13
16	Erect and fix trussed rafter roof, bracings and ties	14 and 15
17	Felt and battens, fascia and soffits	16
18	Lay roof tiles	17
19	Fit uPVC window and door units	17
20	Fit rainwater goods and flashings	18
21	First fix services – cable runs, distribution pipes, etc	17
22	Plasterboard ceilings	21
23	Plaster internal walls	21
24	Second fix services – consumer unit, fittings, fixtures	22 and 23
25	Commission services	24
26	Decorate internally	19 and 24
27	Build paths and driveway	7 and 20
28	Turf lawn	27
29	Clear site	25, 26, 27 and 28

Copy this table and highlight in four different colours the operations that relate to:

- substructure
- superstructure
- services installation
- external works.

Task 7

Look at the final column in the table in Task 6. For each operation in the list, the final column shows the operation (or operations) it must be preceded by. This allows you to determine the sequence of the construction operations that the contractor has planned, based on its resources and preferred method of working as set out in its method statement.

The contractor produces a flow chart of these operations to help organise and monitor the progress of the construction works. A simple flow chart for the Stewards Elm Lane development has been started below. Copy out this figure on the left-hand side of a sheet of A3 paper, then complete the right-hand side of the flow chart showing the remaining construction operations given in the table up to the last operation (29, clear site).

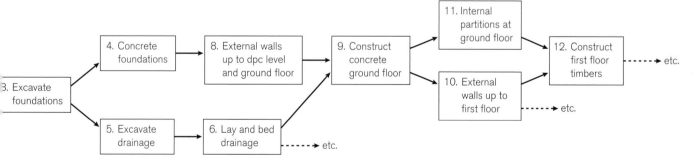

137

ACTIVITY 3

WHAT ARE THE RESOURCES THAT ARE NEEDED TO BUILD A CONSTRUCTION PROJECT?

Task 1

Any construction project requires four basic resources to complete the building phase:

- materials
- plant and equipment
- craft skills (and labourers)
- finance.

In groups, brainstorm to come up with many examples for each of these categories that would needed to build one of the houses in Stewards Elm Lane development project.

Task 2

Consider the three mixed-up lists below, which relate a material, its description and its storage. For each material in list 1, you need to match up the right descriptions in lists 2 and 3. Copy out the table so that the materials, description and storage and disposal are properly collated.

List 1: Material	List 2: Description	List 3: Storage and disposal
Joinery item, eg door architrave	Used typically on doors and windows, such as handles, locks, catches, screws, nuts and bolts.	<ul><li>Nominal size 225 mm × 102.5 mm × 75 mm</li><li>Stacked on pallets off the ground, close to working area</li><li>Often shrunk wrapped</li><li>Waste can be recycled as hardcore or rubble</li></ul>
Plasterboard	A mixture of sand, soda, ash and limestone that is heated in a furnace until molten, which on cooling becomes clear, dense and hard.	<ul><li>Stored in bulk sealed silos for large volumes or in waterproof sacks off the ground on pallets</li><li>Must be kept dry and ventilated at all times</li><li>Sacks used in the order they were delivered to prevent it curing from the moisture in the air</li></ul>
Window glass	Known as commons, facings or engineering, they are made from clay or calcium silicate. They can be held in one hand.	<ul><li>Supplied in various capacity tins which should be stored securely and kept at room temperature</li><li>Waste materials need to be disposed of by specialist waste contractor</li></ul>
Drainage pipes	Component used for the second fix stage of construction including cupboards, doors, windows and stair.	<ul><li>Protect from damage by other trades</li><li>Stored off the ground on closely spaced bearers</li><li>Kept dry</li></ul>
Bricks	Bonded sheet material typically 9.5 mm or 12.5 mm thick with a gypsum plaster core.	<ul><li>Easily dented or punctured</li><li>Stored off the ground on closely spaced bearers to prevent sagging</li><li>Kept dry</li><li>Waste must be disposed of separately</li></ul>

List 1: Material	List 2: Description	List 3: Storage and disposal
Ironmongery	A manufactured material of finely milled and heated clay and limestone or chalk. Forms a grey powder that reacts with water and aggregate to form concrete when mixed.	• Wedged to prevent rolling, stacked horizontally on pallets. • Clay and concrete waste can be recycled as hardcore or rubble
Paint	Rigid cast iron, concrete, clay or flexible uPVC. Used to channel liquid waste products.	• Stored upright in racks • Protected from being smashed or knocked as it can be very brittle • Old materials can be crushed and recycled as 'sand'
Cement	A protective and decorative treatment for wood and some metals, applied in a thin liquid form which dries to a solid film.	• Kept safe from theft in locked containers • Supplied in individual sets, bubble wrapped and boxed to prevent scratches

Task 3

a) Plant comes in three categories:
- small plant is non-mechanical items and tools such as forks, shovels, barrows, ladders and trestles, and mechanical items such as drills, saws and sanders
- large plant includes non-mechanical items as scaffolding, formwork, hoarding and trench supports, and mechanical items such as concrete mixers, transport, cranes, excavators, etc
- administrative and sundry plant is offices, toilets, survey equipment and safety gear.

Using the internet find some images of at least five different types of small plant, large plant and sundry plant. Using your images and brief descriptions create an interesting colour poster.

b) The output of large plant is based on a variety of factors depending on the type of plant. For example, complete this table for an excavating plant (such as a JCB-3CX) that would be used to dig foundations at the Stewards Elm Lane development.

Factors which increase efficiency of task	Factors which decrease efficiency of task
eg the size of the excavation bucket	eg the breakdown time of the plant

Task 4

When contractors tender for a building project, they will carefully study the information that they have been given. This will typically include specifications, drawings, bills of quantities and other contractual information. From this, they will break down the project into a series of individual construction operations and sequence them into a logical order depending on their available resources and expertise. The document that is used to capture this information for the individual operations is the contractor's method statement.

139

The method statement shown below uses typical headings together with sample information for the Stewards Elm Lane project taken from the table in Activity 2, Task 6.

METHOD STATEMENT						
	Column 1	Column 2	Column 3	Column 4	Column 5	Column 6	Column 7
Ref	Activity	Quantity	Method	Output per hour	Labour	Plant	Duration (days)*
1	Site set up	5 cabins/ offices & 1 container	Fix and complete pre-construction elements including temporary services, hoardings etc	-	2 multi-skill trades & 2 labourers	Mobile crane hire	5
2	Strip site and set out	40 m^3 per plot (total 280 m^3 for 7 plots)	Excavate to reduced level over construction area using a tractor-based face shovel; topsoil to be stockpiled on site	40 m^3	Exc. driver, dumper driver, 1 labourer	JCB 3CX backhoe Terex PT6-AWS dumper	1
3	Excavate foundations	27 m^3 per plot (total 189 m^3 for 7 plots)	Excavate foundation trench tractor-based backhoe; surplus removed off site	25 m^3	Exc. driver, lorry driver, 1 banks-person	JCB 3CX backhoe, 10 tonne muck-away lorry	1
4	Concrete foundations	25 m^3 per plot		10 m^3/ hour			
5	Excavate drainage	24 m run per plot		6 m run/ hour			
6	Lay and bed drainage	20 m run of drainage per plot		2.5 m run/ hour			
7	Backfill drainage trenches and compact fill	8 m^3 per plot		16 m^3/day			

* Assuming an 8-hour day

Undertake research using textbooks or the internet to complete the gaps in this statement, that is for construction operations 4–7. Clearly show the materials, labour and plant that are required, and also work out the total duration for each operation, in days, for the whole project.

ACTIVITY 4

WHAT FACTORS CAN CAUSE PROBLEMS WHEN PLANNING AND CARRYING OUT CONSTRUCTION WORKS?

Task 1

a) A building site is a temporary factory, a workshop and a place to store materials. You have to plan carefully when you lay out this temporary factory to make sure that there is an efficient flow of production. This means that you need to think about how building materials are delivered, unloaded and stored prior to use. You must consider how they should be protected against the elements, accidental damage and theft. You also need to consider the movement of workers and plant around the site to ensure that there is easy access and operations do not clash. And this all needs to be done safely.

Write a brief note explaining the factors involved in developing a good site layout. To do this you will need to undertake research. Good sources include *The BTEC National in Construction* by Topliss, Hurst and Skarratt (pages 298–301) and *The Building Construction Handbook* by Chudley and Greeno (pages 84–99).

b) In this task you are going to consider a layout for the Stewards Elm Lane Construction development site shown on page 134. You need to take a copy to work on. First, enlarge it on a photocopier so that it is at least A4 size (though A3 size would be ideal).

Then mark up this plan using coloured pencils to show your proposed site layout in detail. This should include all site offices, entrance and access roads, security hoardings, plant storage compound, temporary service runs, protection for the trees, as well as materials storage for bricks, blocks, roof trusses and scaffolding. Note that the precast concrete floors will be lifted and placed directly from the delivery lorry using a mobile crane.

When you plan where to place the cabins and offices, think about:

- the number of staff working on site (male and female)
- the number and type of site offices units you need
- the need for offices to have a clear view of the site
- the need, ideally, to have a visitors' parking area
- a mess room or canteen (important on larger sites)
- the provision of sufficient toilets and washing facilities
- drying rooms for the workers on site.

For the Stewards Elm Lane site assume that the contractor needs the site accommodation set out in the table below. The sizes of the portable cabins and offices can be found from the Konstructa website (www.konstructa.co.uk).

	Site accommodation	Size /quantity
1	Site office	1 portable office 4.8 m × 2.7 m
2	Security office	1 portable office 3.6 m × 3 m
3	Mess/canteen	1 portable cabin 6 m × 2.7 m
4	Toilets and washing facilities	1 portable cabin 3.6 m × 3 m
5	Changing and drying room	1 portable cabin 4.8 m × 2.7 m
6	Lock-up material and tool store	1 steel container 6 m × 2.4 m

When you plan access to the site, consider:

- checking routes to and from the site to make sure that they are suitable for transporting all plant, equipment and deliveries
- ensuring that there is sufficient space to erect scaffolding and other temporary works
- possibly restricting general movement inside the site, maybe consider a one-way system.
- making access for delivery vehicles and for storing materials near the building being constructed to reduce movement around the site.

For example, consider whether paved areas and roads could be constructed or partially completed at an early stage so that access and movement around the site can be improved. Also consider when and how you will use the materials, so that you can arrange deliveries so that materials do not come all at once. Careful planning of deliveries can reduce damage, theft and unnecessary movement, such as double handling.

Task 2

The planning of construction works must be carried out safely to protect not only the construction workers but also other personnel and the general public.

This activity can be found on the National Learning Network website. All recognised sixth forms and further education colleges are members of the National Learning Network and will have access to these online activities. Your school or college may already have these materials linked to its 'virtual learning intranet' – if not, ask your tutor for the organisational password that you will need this to log into the site.

To access the activity, go to the National Learning Network home page (www.nln. ac.uk). From the home page click through the tabs 'Vocational' then 'Construction and Engineering L3' and then browse through the list of activities to find the one entitled 'Protection of the public'. The activities are listed in alphabetical order.

Alternatively you could carefully type this URL into the address bar of your web browser:

http://go.nln.ac.uk/?req=%7B54F06CA2-06B7-4A74-B40F-42E0C6ACF802%7D&prov=aclProv2264

Work through the activities, writing down your answers in the box provided then checking to see if you are right. Then prepare a draft email that you could send to the site agent setting out your concerns regarding protection of the public.

Task 3

a) The workforce is critical to the success of any construction project. There are several key factors that can severely disrupt a construction programme if the workforce is not managed properly. These are:

- the availability and cost of individual workers
- the workers' skill levels
- the workers' motivation
- the workers' productivity.

Under each of these headings, give one specific example which could positively impact on the Stewards Elm Lane project site and the management of the workforce, and one example which would have a negative impact. Hint: read through a good construction textbook – ask your tutor to recommend one.

b) Study photographs A and B. Describe what each one tells you about the site management's attitude to the workforce on the site and the likely effect on the workers' motivation and productivity.

Photograph A

Photograph B

Task 4

Weather can have a disastrous effect on a construction programme if it is not properly considered at the planning stage. Consider how these operations could be affected by the weather:

- excavating the foundations
- concreting the foundations
- lifting the individual trussed frame rafters into position
- constructing an in situ reinforced concrete retaining wall to a basement.

For each operation, provide an example of how the weather could badly affect this operation at the Stewards Elm Lane development, and suggest what could be done at the planning stage to reduce the risk of delay due to weather.

Task 5

Material shortages occur when manufacturers face an unusually high demand. For example, after the 2007 gales when everybody's fences were blown down there was a shortage of fencing panels available for sale. Demand was so high that contractors were having to scour the country for timber fence panels.

It is therefore important for a contractor to provide material manufacturers with details of what quantities they will require (and by what dates) as soon as the contract is signed. This should allow for any lead-in time for fabrication or procurement of the materials, so that they will be available well in advance of the delivery date to site.

Contact some local builders' merchants or visit appropriate websites to source product literature and manufacturers' catalogues. Using this information, determine suitable alternative specifications for these materials and components and, where applicable, state typical lead-in times that are needed for their production:

a) hardwood double-glazed patio door

b) granite sets for an external path and driveway

c) genuine slate roof tiles

d) marble floor tiles for the bathroom.

Task 6

These three scenarios describe some projects for typical clients.

1. A young family on a tight budget wants a small three-bedroom house built. A family relative, who is a mechanical engineer, is drawing up the plans for local authority approval.

2. A wealthy middle-aged couple want a second home on the south coast for holidays. They have employed a chartered architect to design and project manage their 'grand design' using the latest technology.

3. A commercial pub chain wishes to refit a high street restaurant. It has an in-house architectural team that undertakes all its planning drawings and approvals.

Suppose you are a construction contractor. You are considering whether to tender for each of these projects. For each one, assess the likelihood of late design changes and the possibility of late construction information. Explain clearly the judgements that you make and any assumptions that you have made.

ACTIVITY 5

CONSTRUCTION MANAGEMENT PROCESSES

Task 1

Undertake suitable research on the internet, or from a recommended textbook, to find as many examples as you can of each of these documents.

Labour documentation
- Time sheets and/or job cards
- Wages sheet
- On-costs/overheads
- Bonus payments/performance-related payments
- Performance figures
- Work instructions
- Employment conditions

Materials documentation
- Orders and requisitions
- Delivery tickets and/or goods received sheets
- Invoices
- Stock control documents
- Waste control documents

Plant documentation
- Invoices
- Usage rates
- Vehicle allocation sheets
- Utilisation rates or idle time records
- Maintenance schedules

Progress and finance documentation
- Gantt charts, precedence diagrams and critical path network analysis
- Payment claims
- Subcontractor invoices/payments
- Payments received
- Cashflow forecasts
- Diary/meeting minutes

Task 2

Using your research for Task 1 above, attempt to answer these questions.
Produce your answers in written paragraphs.

1. Why is important for all operatives to complete a time sheet whether they are subcontract labour or directly employed?
2. Give some examples of typical overhead costs that a main contractor has to cover. How can the contractor monitor these costs?
3. Bonus payments are often used to incentivise the workforce. What precautions need to be considered when setting up bonus schemes?
4. What clauses in a worker's employment contract can help the contractor (employer) prevent pilfering of materials and tools from the site? And how can this be monitored and enforced?
5. With regard to material deliveries, what is the importance of delivery tickets? What actions should be taken before a delivery is signed off? Which of the documents listed in Task 1 records the materials that are stored on site at any given time? Why is this useful to the site manager?
6. Which documents give an indication of efficient use of materials? And why is this financially and environmentally important?
7. List three advantages of buying plant and three advantages of hiring plant.
8. Why is cashflow so important to the contractor and which documents above contribute to working out the cashflow?
9. Why are site meetings so important for keeping the project on programme and on budget?

ANSWERS

ACTIVITY 1

Tasks 1, 2 and 3

Role	Designation and function
Director or chief executive	P, Pl, R
Contracts manager	P, Pl, R
Site manager	T/P, Pl, R
Estimator	T, Pl, R
Buyer	T, R
Planner	T, Pl
Cost control surveyor	T/P, Pl, R
Health and safety consultant/officer	P, Pl
Document controller	S, R
Subcontractor (manager)	T/P, Pl, R (and Pr)
General foreperson	S, R, Pr
Craft operative	C, Pr
Manual operative	L, Pr

Key:
P	Professional
T	Technical
S	Supervisory
C	Craft
L	Labourer
Pl	Construction planning
R	Resource management
Pr	Construction production

Task 4

a) Marcus Hutching is the architect for the project and, under the RIBA Plan of Work, his role is to co-ordinate and lead the design team. He also ensures that the client's requirements are met by the contractor by chairing the weekly site meetings.

b) Bestend is represented by three levels of its organisation at the meeting. The highest ranking is the contracts manager, Jack Barrymore. His role is to oversee the allocation of resources between different construction projects and to act as a link between the projects on site and the senior management based at head office. The person who is responsible for all the construction works for this site is Steve Jones, the site manager. He is the main person responsible for ensuring that all the resources are in place for the project to be built safely and to the architect's specification. Rebecca Donovan is responsible to Steve Jones for assisting in programming the works and in carrying out the valuations of the work that has been done.

c) Alf Richards is the site-based supervisor responsible for the Dufflin groundworkers. He would report to his boss Paul Hudson on the technical and labour problems that have been encountered, and also possibly make suggestions on the best solution to remove the obstructions. Paul Hudson is Dufflin's managing director who would lead at the meeting to discuss whether his company can implement a solution to the problem. He would also provide approximate costs in time and labour for the additional work.

d) WRP are the consultant engineers on the project and, as part of the design team, will have undertaken or have organised the initial site and soil investigation. At the meeting they will need to reassure the client through their representative Hassam Chowdary that they undertook all appropriate measures to identify any underground obstacles. They will also need to advise the architect on whether parts of the new foundations will need to be redesigned if the obstacle cannot be removed.

e) The person directly responsible for the overall programme of works and ensuring that all the resources are in place to build the development is Steve Jones, the site manager. He needs to be certain that any delays with the foundations can be 'made up' by reviewing the programme for some of the activities that follow the foundation work. He will ask Rebecca Donovan, his planner, if there is any 'float' or spare time available that could ease any possible delay, and hence ensure that the project completion date is still met. Bestend Contracts would compare the alternative solutions based on their effects on the overall construction programme.

f) The factors that will govern the choice of solution to this unexpected problem are:

- the cost and time of any additional demolition/excavation plant that may need to be hired, including all the necessary transport costs – to be investigated by Bestend Contracts and Dufflin

- the cost/time of redesigning the new foundations and any additional construction labour and material costs that may be incurred – to be investigated by WRP Engineers and Dufflin

- the availability of Dufflin's workforce to undertake any additional groundworks given any other projects that the company is already committed to – to be investigated by Dufflin.

g) Bestend Contracts is the main contractor with the responsibility to build the project, and the final decision on how to overcome this problem will be its responsibility. The contracts manager Jack Barrymore, in consultation with the senior management, will probably review the alternative solutions and make the final choice. Representatives from the architects, Rackman Design Partnership, will want to be assured that the original scheme is not going to be compromised and will be built according to its client's requirements, but this firm will have no say in how the works will be constructed.

Task 5

Marcus Hutching – the architect
Communicates: with client; with all design team members; with main contractor.
Co-operates: with design team when seeking design solutions.
Instructs: design team and contractor when there are design changes to be made.
Co-ordinates: the design development and approval; the selection of the contractor; the site progress meetings.
Controls: the design process.

Jack Barrymore – Bestend's contracts manager
Communicates: with head office; with the architect; with the site manager.
Co-operates: with the architect and client.
Instructs: the site manger overseeing site work.
Co-ordinates: the deployment of resources between construction projects.
Controls: the staffing and materials resources between construction projects.

Steve Jones – Bestend's site manager
Communicates: with all site-based staff including subcontractors; with design team.
Co-operates: with architect and design team.
Instructs: directly employed labour; all subcontractors; material and equipment suppliers.
Co-ordinates: the deployment of resources on site.
Controls: all site operations via site supervisors.

Rebecca Donovan – Bestend's planner and cost control surveyor

Communicates: with the site manager; with the architect/QS on cost issues.
Co-operates: with the construction team.
Co-ordinates: the construction programme; payment of interim valuations.

Paul Hudson – Principal of Dufflin

Communicates: with architect; with site manager; with his site operatives.
Co-operates: with site manager and architect.
Instructs: his groundwork foreman.
Co-ordinates: the deployment of resources between various projects.
Controls: the staffing and materials resources of the groundwork company.

Alf Richards – groundwork foreman for Dufflin

Communicates: with principal of Dufflin; with site manager.
Co-operates: with site manager.
Instructs: groundwork operatives.
Co-ordinates: the day-to-day work of the groundwork team.

Hassam Chowdary – consultant structural/civil engineer working for WRP Consultants

Communicates: with architect; with other design team members.
Co-operates: with design team when seeking design solutions.
Instructs: contractor where there are design changes to be made.

ACTIVITY 2

Tasks 1 and 2

Order	Title	Description
1	Client's brief	E
2	Sketch designs	D
3	Final scheme	G
4	Tendering for a contractor	H
5	Awarding the contract	B
6	Plan and mobilise for construction	K
7	Construction operations	C .
8	Defects period and final completion	A
9	Maintenance	J
10	Refurbishment and renewal	M
11	Decommissioning	L
12	Brownfield site redevelopment	F

Task 3

Feasibility stage: the initial formulation of the client's requirements, the identification of the site and the site constraints.

Design stages: the formulation of the design ideas into a design scheme that satisfies the client and the statutory approval bodies such as the local authority planning department.

Pre-construction stages: the planning of the construction works and mobilisation of all the supporting facilities such as welfare and temporary services etc. It will also include ordering the materials and subcontract labour and contracting the manufacturers to produce any prefabricated components

Construction stages: the organisation, co-ordination, monitoring and controlling of site operations to ensure that the project is built on time, on budget and to the correct specification.

Review stage: receiving feedback on whether the outcomes have been successfully achieved for the project. Examining the performance of individuals and companies; using this information in the planning and tendering of future projects.

Task 4

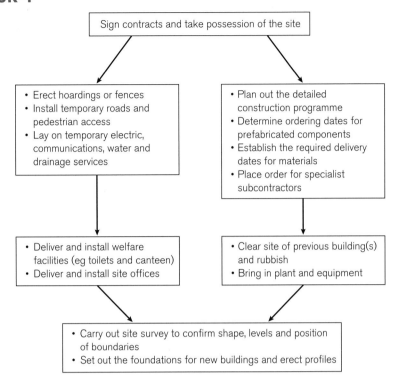

Task 5

There are no specific answers to this activity – refer to answers on National Learning Network website.

Task 6

Lines 2–8 should have been highlighted together as substructure operation.

Lines 9–20, 22, 23 and 26 should have been highlighted together as superstructure operations.

Lines 21, 24 and 25 should have been highlighted together as services installation operations.

Lines 1 and 27–29 should have been highlighted together as external works operations.

Task 7

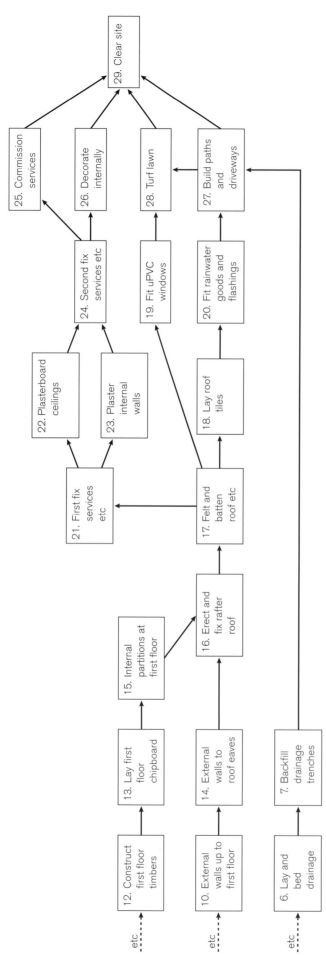

ACTIVITY 3

Task 1

There are no specific answers to this task.

Task 2

The correct matching of is as follows:

List 1: Material	List 2: Description	List 3: Storage and disposal
Joinery item, eg door architrave	Component used for the second fix stage of construction including cupboards, doors, windows and stairs.	• Protect from damage by other trades • Stored off the ground on closely spaced bearers • Kept dry
Plasterboard	Bonded sheet material typically 9.5 mm or 12.5 mm thick with a gypsum plaster core.	• Easily dented or punctured • Stored off the ground on closely spaced bearers to prevent sagging • Kept dry • Waste must be disposed of separately
Window glass	A mixture of sand, soda, ash and limestone that is heated in a furnace until molten, which on cooling becomes clear, dense and hard.	• Stored upright in racks • Protected from being smashed or knocked as it can be very brittle • Old materials can be crushed and recycled as 'sand'
Drainage pipes	Rigid cast iron, concrete, clay or flexible uPVC. Used to channel liquid waste products.	• Wedged to prevent rolling, stacked horizontally on pallets. • Clay and concrete waste can be recycled as hardcore or rubble
Bricks	Known as commons, facings or engineering they are made from clay or calcium silicate. They can be held in one hand.	• Nominal size 225 mm × 102.5 mm × 75 mm • Stacked on pallets off the ground, close to working area • Often shrunk wrapped • Waste can be recycled as hardcore or rubble
Ironmongery	Used typically on doors and windows, such as handles, locks, catches, screws, nuts and bolts.	• Kept safe from theft in locked containers • Supplied in individual sets, bubble wrapped and boxed to prevent scratches
Paint	A protective and decorative treatment for wood and some metals, applied in a thin liquid form which dries to a solid film.	• Supplied in various capacity tins which should be stored securely and kept at room temperature • Waste materials need to be disposed of by specialist waste contractor
Cement	A manufactured material of finely milled and heated clay and limestone or chalk. Forms a grey powder that reacts with water and aggregate to form concrete when mixed.	• Stored in bulk sealed silos for large volumes or in waterproof sacks off the ground on pallets • Must be kept dry and ventilated at all times • Sacks used in the order they were delivered to prevent it curing from the moisture in the air

Task 3

a) There are no specific answers to this part of the task

b) These are some suggestions regarding the efficiency of the excavation plant:

Factors which increase efficiency of task	Factors which decrease efficiency of task
• The torque or power of the machine • The traction of tracks on the subsoil surface • The size of the excavation bucket • The number of lorries available to be loaded • The skill and experience of the operator	• Idle time waiting for the lorry to return from the tip • The swelling or bulking of the soil when it is excavated • The breakdown time of the plant • The idle time for individual operators' rest times and comfort breaks

Task 4

METHOD STATEMENT						
	Column 1	Column 2	Column 3	Column 4	Column 5	Column 6	Column 7
Ref	Activity	Quantity	Method	Output per hour	Labour	Plant	Duration (days)*
4	Concrete foundations	25 m³ per plot	Concrete to be ready mixed and delivered to site and chute loaded direct into foundation trench and compacted with vibrating poker	10 m³/hour	1 banksperson and 3 operatives in concreting gang	6 m³ ready-mix lorry; 2 × pokers; compressor	3 plots per day
5	Excavate drainage	24 m run per plot	Excavate drain trench with tractor-based backhoe to indicate levels, and stockpile alongside trench	6 m run/hour	1 driver and 1 banksperson	JCB 3CX backhoe	0.5 plot per day
6	Lay and bed drainage	20 m run of drainage per plot	Bed and lay drains with pea-shingle and lay ic and runs	2.5 m run/hour	2 operatives	1 pipe layer, automatic level, shovels	1 plot per day
7	Backfill drainage trenches and compact fill	8 m³ per plot	Backfill with excavated soil and compact in 300 mm layers	16 m³/day	1 driver and 1 banksperson	JCB 3CX backhoe, whacker plate	0.5 plot per day

* Assuming an 8-hour day

ACTIVITY 4

Task 1

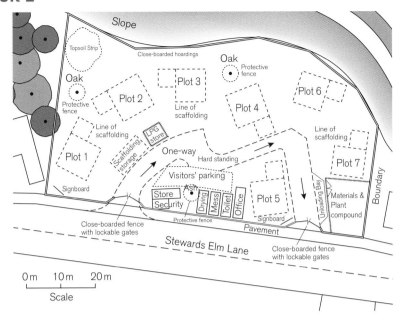

Task 2

There are no specific answers to this task.

Task 3

a)

Workforce factor	Negative example	Positive example
The availability and cost of individual workers	When the economy is booming there is a plentiful supply of work so there may not be enough labour to go around and consequently the hourly rates of workers will increase, which will make the job more expensive and reduce the contractor's cashflow.	When the economy is declining, there is a glut of labour looking for work so the contractor can pay lower rates and keep costs down.
The workers' skill levels	In certain trade skills there is not a sufficient quantity of labour to meet the demand. Therefore certain projects have to wait until a specialist is available. This often applies to construction craft operatives that work on heritage buildings where a greater level of skill is required.	With the increasing use of prefabrication and offsite construction, workers with general, low skills can be quickly trained up to learn within a factory-controlled environment. There is also the increasing availability of multiskilled craft operatives who are able to maximise their usefulness, to the mutual benefit of themselves and their employer.
The workers' motivation	Subcontractors can be offered bonuses. If this is done on a piecework basis – say workers are paid for the number of components that they install – there must be careful monitoring of the work to ensure that the quality of the work is maintained.	Workers can be offered bonus schemes such as a company vehicle or private healthcare. However, simple things such as clean, dry and pleasant welfare facilities with, say, a television in the workers' canteen can improve motivation.
The productivity of the workers	Where work is delayed in one area the site management must ensure that it can make that time up elsewhere in the programme. This may mean flooding the future tasks with additional plant and/or labour. However, depending on the site accessibility and working area, this may cause congestion problems and more delay.	The rates at which workers operate need careful study. This is called work and method study and it involves recording the output per time period for each activity undertaken. It is particularly useful where new products or plant are being used as future jobs can be planned more efficiently and there is less wasted time.

b) Photograph A shows poor site layout and welfare facilities. There are no defined walkways for pedestrians, with separation from vehicular traffic. The surface is uneven which could lead to slips and falls occurring. There seems to be a lack of organised materials storage.

Site workers are unlikely to be well motivated when faced with this sort of working environment. They will probably be fearful for their safety and there could also be legitimate claims for injury and accidents against the main contractor.

Photograph B shows a well-organised site with clear pedestrian access. The materials are stored in an organised manner

The workforce will be confident about the site management and should be well motivated and content. A tidy well-kept site gives a good purposeful working impression to both site staff and visitors alike. The company is also less likely to incur accident claims.

Task 4

Weather can have a disastrous effect on a construction programme if it is not properly considered at the planning stage.

Excavating the foundations
Storms could fill up foundation trenches, which would cause the sides of the excavation to become unstable. This will need temporary walings/poling boards and sump pumps installed in the bottom of the excavation.

Concreting the foundations
Wet concrete cannot be placed and compacted when the temperature is less than 2°C and falling as it will not cure or set properly. It is important to keep concrete covered with insulating mats or enclose and heat artificially.

Lifting the individual trussed frame rafters into position
In windy conditions, with speeds in excess of 20 mph, lifting can be difficult. Needs careful preplanning to avoid abortive crane hire.

Constructing an in situ reinforced concrete retaining wall to a basement
Temperature can be an issue when concreting, as can the presence of excess groundwater. To reduce groundwater issues, install temporary ground dewatering system, such as well pointing or the use of chemical grouts.

Task 5

There are no specific answers to this task.

Task 6

1. There may be late changes owing to inexperience of the building designer and issues with achieving statutory approval. However, the form of construction is relatively standard and the house should be straightforward to build. The tight budget will have a tendency to restrict any changes.

2. The clients will want to have a large input in the specification of their ideal home and although they are employing a qualified architect it does not mean that they will not change their minds. Another issue is that with using the 'latest technology', there may be unexpected technical changes if not properly researched in advance. New methods and unfamiliar processes may have to be used by the contractor – and this may not run as smoothly as using traditional and known methods.

3. The client is a commercial company that should be fully aware of its budget and the quality of the job required. The close relationship between the 'client' and 'designer' should ensure that the work is planned and organised well, leaving the contractor to carry out the works without major design

or construction changes. However, one point to be aware of is that when refurbishing old existing buildings there may be unforeseen issues with the condition of the property as the old finishes are removed. These may cause a later rethink by the client and/or designer, as well as damp-proofing and structural problems.

ACTIVITY 5

Task 1

There are no specific answers to this task.

Task 2

1. It is important for all operatives to complete a time sheet whether they are subcontract labour or directly employed because a close tally can then be kept on the amount of time taken to undertake specific site activities. This information will be useful for future estimates. It also provides a check on the financial outgoings of the site and verifies that work is being claimed legitimately. Many modern site offices have electronic clocking in facilities that rely on fingerprint recognition so the process can be integrated within management and financial systems.

2. Typical overhead costs that a main contractor has to cover are head office costs, such as power, lighting and service charges. There are salary costs for administrative staff such as personnel and accounts. Telecommunications, specialist software and internet access are also important overheads that need to be accounted for. Overheads must be monitored using auditing procedures to check invoices against orders to avoid fraud and to reduce waste. It is important to measure the performance of off-site operations as much as on-site operations. It is only by measurement that improvements can be made and reviewed.

3. Bonus schemes are often used to improve productivity for site work. The schemes that are used must meet the demands of the project. If finishing the project within a certain time is a priority then schemes which count the number of elements completed is important, but care must be taken to ensure that the work meets the required quality. Piecework often leads to contracts finishing on time but the quality of work may not be to the correct specification. Money is not the only incentive. Some companies offer shares in the company or other linked products as well as access to services such as loyalty cards or company cars.

    Targets for bonuses must be adequately gauged to ensure that workers are being treated fairly. If the target is set too low, then there is no incentive. However, if it is set too high it may be impossible to achieve the target. In all cases it is advisable to carry out suitable 'work study' measurements prior to setting up the bonus scheme to ensure that rates are set fairly.

4. All workers under contract are required to let their vehicles be searched, when leaving the site, by the designated managing agent. In this way, workers can be prevented from pilfering tools and materials from the site. To implement this policy, there must be specially trained security site staff.

5. All material deliveries to the site must be checked by a competent person to ensure that the correct quantity has been supplied and that it is of the right quality. The delivery ticket is signed to show what has been received; this information is then recorded on a goods received sheet which is then matched against the supplier's invoice for payment to be made. The goods received sheet helps the contractor know that the work is progressing and that materials are available. In the same way, plant hire sheets are used to track and monitor the use of the construction plant.

6. The document that gives an indication of efficient use of materials and labour is the operative's job card. This is used to monitor the actual hours against the planned hours. These can then be used to help set the correct bonus rates for directly employed workers. The documentation of the removal of waste from site is also important to ensure that the landfill costs are monitored and that waste materials are minimised.

7. Buying plant is advantageous when the plant is used regularly and often, and for keeping on/off costs to a minimum. It can work out cheaper if the plant is maintained regularly. Also, owned plant tends to be used more carefully by the operators.

    Hiring plant is useful where specialist 'one-off' equipment is needed such as cranes or pile rigs. If the equipment breaks down, then replacements can be supplied at no extra cost (or delay) and the expense is carried by the hirer. Some large plant has to be hired with a driver.

8. Cashflow is so important to the contractor because this is the means by which the day-to-day running of a project is financed. The cashflow within the project is managed by the contractor's quantity surveyor who undertakes interim valuations on the work that has been done, in agreement with the client.

9. Site meetings are very important for keeping the project on programme and on budget. The main contractor hosts a weekly meeting on site and invites all the key professionals and trades. It is an opportunity to receive feedback on the progress of the works, agree courses of action for current issues and to plan (and communicate information about) future works.

MARKED ASSIGNMENTS

UNIT 1 – HEALTH, SAFETY AND WELFARE IN CONSTRUCTION AND THE BUILT ENVIRONMENT

This assignment is about the legal responsibilities of both employers and employees, and accident recording and reporting procedures. It will lead to the demonstration of the following learning outcomes from the unit:

1   Understand the general and specific responsibilities both of employers and employees under current health, safety and welfare legislation

2   Be able to identify workplace hazards, persons who may be affected by such hazards, and the potential consequences of accidents

Content

Health, safety and welfare are of vital importance to any company working within the construction industry. The Health and Safety Executive's statistics of fatal injuries shows the construction industry consistently at the top the table. To try and improve the situation a framework of UK and European Union legislation has been created to prosecute employers and employees who disregard health, safety and welfare.

This assignment requires you to show you understand the legislation that applies to health, safety and welfare and the roles of, and interaction between, the key personnel. You will also need to demonstrate knowledge of an individual's responsibility in accident recording and reporting procedures as a method of improving health, safety and welfare.

This assignment will enable you to provide the evidence required to meet the following grading criteria:

P1   *Identify and describe the roles and responsibilities of the persons responsible for health, safety and welfare on a construction project*

P2   *Identify three main pieces of health, safety and welfare legislation relevant to the construction and built environment sector and describe the legal duties of employees and employers in terms of such legislation*

P6   *Identify and describe the role of the individual in accident recording and reporting procedures*

M3   *Explain how collecting accurate data and information on accidents and incidents contributes to improvements in health, safety and welfare in the workplace*

M1   *Explain how the members of the building team interact in terms of their health, safety and welfare roles and responsibilities*

D2   *Evaluate a provided accident report and suggest improvements that could be made to workplace systems in the future to avoid a recurrence*

You will need to reference your sources and provide a bibliography.

Task 1 (P1)

Identify and describe the roles and responsibilities of the persons responsible for health, safety and welfare on a construction project

Describe the specific health and safety roles and responsibilities of each member of a building team. Pay particular attention to the formal roles of the client, the main contractors, subcontractors, the HSE, the local authority and the planning co-ordinator.

Task 2 (M1)

Explain how the members of the building team interact in terms of their health, safety and welfare roles and responsibilities

Write an account to identify and explain how the building team members interact with each other in terms of their health and safety roles and responsibilities. Consider these questions.

- Who reports to whom?
- Who is responsible for other members of the team?
- Who does what?

Task 3 (P2)

Identify three main pieces of health, safety and welfare legislation relevant to the construction and built environment sector and describe the legal duties of employees and employers in terms of such legislation

State why each piece of legislation is relevant, and what it controls. Explain (with examples) what the duties of the employer **and** the employee are in respect of each piece of legislation.

Task 4 (P6)

Identify and describe the role of the individual in accident recording and reporting procedures

Imagine that there has been an accident on site. Use the HSE website to:

a) find a small case study of an accident that has occurred on a construction site – write your own overview of these events in your leaflet

b) identify and describe the role of the individual in recording this accident and the procedures for reporting it to a supervisor or manager – consider the procedures for reporting dangerous occurrences and near misses too.

Task 5 (M3)

Explain how collecting accurate data and information on accidents and incidents contributes to improvements in health, safety and welfare in the workplace

In your place of work there is quite a rigorous procedure in place to collect and analyse the company's accident data. The selection task requires you to prove why this is good practice.

a) Explain how collecting accurate data and information on accidents and incidents can contribute to improvements in health, safety and welfare in the workplace. Demonstrate how an analysis of results forces change.

b) Use health and safety statistics to explain your findings.

c) Discuss relevant issues such as recording incidents by type of injury (minor, major or fatal), location, cause of accident, together with compiling data on those involved (by gender, age, occupation, etc).

Task 6 (D2)

Evaluate a provided accident report and suggest improvements that could be made to workplace systems in the future to avoid a recurrence

You have been asked to evaluate this accident report and suggest improvements that could be made to the workplace systems in the future to avoid a recurrence.

Accident report

Date: Tuesday 13 May

Employee: J Brownsmith

Location: Stewards Elm Lane construction site

Description of event:

At about 4.30 pm on Tuesday 13 May I was locking up the construction site compound and walking towards the site entrance gate when I fell into an excavation and injured myself on the reinforcing bars.

I was on my own and had to wait until the visiting security guard found me and helped me out of the excavation.

I was taken by ambulance to the hospital to have a cut to my leg stitched and dressed. I reported to work the next day as normal.

TASK 1 ANSWER

Unit 1 Assignment 1 Task 1 (P1)
Name: Jo Hussein
Identify and describe the roles and responsibilities of the persons responsible for health, safety and welfare on a construction project

I am going to discuss the roles and responsibilities of the key personnel in the office development.

Client

The client is the individual or organisation who needs building work. Clients build for two reasons; either for themselves or for other individuals or organisations. The client provides the original brief and finances the project, they also agree to make payments on time.

The health and safety responsibilities of the client involve allowing sufficient time for the project to be completed safely. They must appoint a CDM co-ordinator (CDMC) and a principal contractor, also when appointing people, they are to make sure that only competent and adequately resourced people are employed. Clients are also responsible for organising that a satisfactory Health and Safety plan *is* written before construction *starts* and ensure there is a Health and Safety File available for inspection, but this responsibility is usually given to the CDMC to organise.

CDM Co-ordinator (CDMC)

Some of the roles and responsibilities of the CDMC involve advising the client throughout the whole project, producing a Health and Safety plan, and making sure that there is enough time to complete the project safely, and that all parties are working to the CDM regulations.

Designer

The designer's role is to act as the client's agent; they convey the client's needs into a feasible design brief. Other responsibilities of the designer include obtaining planning permission for the client, constructing a site investigation; desk study and a walk-over survey. Also he/she studies the circulation and space requirements and identifies constraints. As the work proceeds it is up to the design team to ensure that the contractor is provided with information sufficiently in advance to enable him to proceed without causing any delay to the works. They must design buildings which can be built safely without putting the contractor at unnecessary risk. They must think about safety when they specify certain materials and components as some may be too heavy or give off toxic fumes. They must also build into their design how the building will be safely used throughout its life like for instance how the windows can be safely cleaned.

Quantity Surveyor (QS)

The quantity surveyor is the client's cost consultant. The cost of a project must be controlled at all stages. On some contracts the quantity surveyor produces a detailed bill of quantities which lists, in a standard format, the work to be done and the materials that are required. The quantity surveyor measures the work done by the contractor as it proceeds and compares it with the drawings noting any changes. The value of the work which has been satisfactorily completed can then be calculated. Then he/she draws up the interim payments to the contractor and prepares the final account.

Main Contractor

The main contractor carries out the construction of the building in accordance with the architects drawings. When the contractor starts on site there must be working drawings available to enable commence, and to facilitate understanding

Please make your introduction more descriptive and in context. Are you concentrating on the formal responsibilities or site work in general?

Good – 'competency' is very important in health and safety management.

Clients do not 'organise' an H&S plan but they must ensure that one is produced by the CDMC.

This is a good example, Jo. Designers need to appreciate how the building will be maintained and design safe systems of work to enable the maintenance to be carried out.

But what are the QS's responsibilities towards health and safety? How do they help the client and designer to create a safe design?

of the size, shape and nature of the structure. It is the contractor's responsibility to complete the works on time, in relation to the contract documents such as the drawings and specifications.

The main contractor provides a Health and Safety procedure and policy for the works. He or she has to ensure that both procedure and policies that are put in place are followed by himself and others. He or she has to ensure that risk assessments and method statements are also being done and safe systems of work are in place. The contractor must also provide welfare facilities for the site like toilets and washrooms. Finally, the contractor should provide health & safety training for all the workers – like tool box talks.

The 'main contractor' in health and safety terms is usually referred to as the 'principal contractor' and the subcontractors as a 'contractor'.

Subcontractor

Subcontractors are all the site workers and include craftsman and labourers. They must each ensure that they have a clear and good understanding of health and safety in the work place. They are responsible for their own and their employees health and safety. They must wear all the proper personal protective equipment; obey all the H&S instruction and signage. They must use safe systems of work and work in a safe way e.g. using scissor lifts instead of step ladders – as they do not provide a safe working platform with handrails. They will also produce risk assessments and method statements for their work. Also it is their responsibility to report any hazards to the main contractor.

Some site workers may be directly employed by the principal contractor but they too must be aware of their H&S responsibilities.

Site Visitors

All authorised site visitors like sales reps also have a duty to obey all H&S signs and instructions given by the management. They must also wear any PPE that they are given and not wander off by themselves.

The HSE Inspector

HSE stands for the Health and Safety Executive. They are employed by the government to check that the health and safety regulations are not being broken on site. They can stop construction works if there are dangerous practices on site like working on a scaffold without handrails or toe boards. Or they can force contractors to change their method of working if they feel that it is unsafe. This is called issuing an improvement notice.

ASSESSOR FEEDBACK FORM

STUDENT GRADING SHEET
Name: Jo Hussein
Unit 1 Assignment 1 Task 1 (P1)

Some satisfactory work done here, Jo.

You have identified the main participants involved in health and safety management and provided a useful snapshot of their formal roles. You have also shown an awareness of the site rules that all site operatives and visitors must meet.

The CDMC has a pivotal role in making sure that H&S planning for the project is carried through from the design stage and then on to construction. You could also mention the role of the local authority, which is responsible for ensuring environmental nuisance from construction sites is avoided.

In future, include the references that you used for the task.

Grading criterion P1 met.

TASK 2 ANSWER

Unit 1 Assignment 1 Task 2 (M1)
Name: Jo Hussein
Explain how the members of the building team interact in terms of their health, safety and welfare roles and responsibilities

The basic hierarchy of responsibilities between the key personnel are summarised in the diagram below.

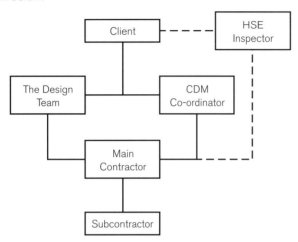

Be careful! The HSE has legal authority over the client, it is not equal in authority!

The client appoints the CDM Co-ordinator and the Designer and he has to check that they are both competent. The main contractor will take advice from the CDM Co-ordinator (CDMC) on health & safety as well as being instructed by the designer about the design. The main contractor must make sure that he passes on all the right health & safety information to his workers and the other subcontractors. If there are any problems on site the main contractor can ask the CDM co-ordinator for help. The subcontractors are the last in line to get help from the CDMC.

The designer team includes the architect, the quantity surveyor and the specialist engineers. They have many roles to play, but are also involved in health and safety. The designers must try and eliminate possible risks and dangers in their design. This helps when constructing so that contractors have a easier job with less risk to injury or accident. This could be things like making objects lighter so that they can be lifted easily, or like trying not to design anything near power cables or other potential risks. It is their responsibility to design things with safety in mind and it is the CDMC who needs to check that this is the case.

The designers must also work with the client to inform the client of any responsibilities at the earliest possible stages. The client must also up date the CDMC of any changes or other information that is an importance to know. The designer could be seen as responsible for those who are carrying out the work. This can be understood by if the designer does not comply with health and safety plan and does not review possible harms or doesn't try and reduce the risks; they could then be liable if anyone injures themselves on a risk that should have been amended.

The contractor has the responsibility of communicating with the designer about current issues which might affect safety hazards which are preventing work commencing. The contractor must also take responsibility of their hired team and ensure they are all competent and safe workers. At planning stages the contractor must also develop the current health and safety plan with help from the CDMC using information provided by the design team. This helps get more health and safety hazards out of the way and help keep the contractors more safe and aware of current risks. Furthermore the contractor must co-operate with the CDMC to help create the success of new health and safety rules for the file.

The CDMC is not directly responsible for site workers. CDMCs assist in the planning and management of safe systems of work but have no authority over the workforce. Do not refer to roles as either he or she; keep them neutral.

Give an example – such as the uncovering of contaminated fill materials in the soil when excavating foundations.

The HSE Inspector is responsible for investigating accidents and prosecuting all those who have broken health & safety laws. He can come onto a construction site at any time without notice as they have a 'statutory right of entry' and can ask to see any parts of the works that he thinks or has been reported as being unsafe. He can stop the work site by issuing a prohibition notice. He can also prosecute the architect and client or any one who has responsibilities for health & safety.

Very good point, Jo, regarding the right of access.

ASSESSOR FEEDBACK FORM

STUDENT GRADING SHEET

Name: Jo Hussein
Unit 1 Assignment 1 Task 2 (M1)

Jo, your response here has been a little vague. You needed to tie down the responsibilities of each person to what they do and who they report to. Your hierarchical chart was a good idea but has over-simplified the relationships, particularly in how the principle contractor and designer respond to the CDM co-ordinator. Also the relationship between the principal contractor and subcontractors roles need to be clarified — it would be good to include some specific case study examples that you researched in class.

Again, please reference your sources.

Grading criterion M1 not yet fully achieved.

TASK 3 ANSWER

Unit 1 Assignment 1 Task 3 (P2)
Name: Jo Hussein
Identify three main pieces of health, safety and welfare legislation relevant to the construction and built environment sector, and describe the legal duties of employees and employers in terms of such legislation

I am now going to discuss three main pieces of legislation for this task.

The Health and Safety at Work Act (1974)

HASW A (the Health and Safety at Work Act 1974) is one of the three main pieces of legislation regarding health and safety. This is to ensure that everybody is responsible and aware of health and safety in the workplace. It is set up to make people more aware of health and safety and ensure the working environment is a safe place to be. The HASWA is divided up into eight sections; each section concentrating on a different aspect of health and safety.

The health and safety at work act 1974 requires the employer to ensure, as far is reasonable practicable, the health and safety of their employees while they are at work. This indicates that measures must be taken into consideration concerning how much is known about a hazard. They must also take into consideration the severity of any harm or injury that may occur from a hazard, and the likelihood of that harm or injury occurring. The employer must also take in mind availability, suitability and the cost of removing or controlling a hazard.

'Reasonably practicable' is a key term in the management of health and safety.

Employers

The employer has specific legal responsibilities these are known as duties of care. These duties are that the employer must inform the Health and Safety Executive (HSE) regarding construction operations expected to last for the duration of six weeks or more. All of the employer's plant must be maintained in good working order. To avoid risk to employees, materials should be handled and stored correctly, this is the responsibility of the employer. The employer must ensure that their site/workplace is in a safe condition and that this is always maintained. A safe working environment must be created and maintained regarding such hazards as noise and dust.

Employers have duties to provide adequate and appropriate training in safe working practices to all staff and must also provide staff with adequate and appropriate personal protective equipment (PPE) required at no extra cost. Staff and operations must also be given necessary information to allow them to perform their work safely and provide correct supervision.

Employers also have a care of duty of safe working conditions for people not directly employed by them self, this would include such people as subcontractors. A safety committee should be set up and maintained by the employer, this will ensure adequate communication throughout the work force. The employer should also communicate with the employees regarding hazards this could be done using a video or tool box talks.

Employers also have a duty to the public for safety. This includes trespassers, children and thieves, even though the people should not be on the site and may be breaking the law doing so.

The safety of the public is very important and often overlooked!

Employees

Employees also have responsibility in maintaining their own health and safety and the health and safety of colleagues and the general public. The employee must take reasonable care not to put themselves at risk. The employee must not intentionally damage or interfere with machinery, plant or equipment, and must not obstruct any inspector during there execution of their duties. It is the employee's responsibility to pass on relevant information relating to safe working practice. The employee must also ensure that PPE provided by the employer is used correctly. Employees must report to their supervisor about defective equipment on site. The employees must be provided with supervision if they are new staff or are young such as an apprentice or trainee. Employees must ensure that they co-operate at all times with the employer in relation to the stated safety policy. Self-employed companies have to follow the same regulations as employers and the employees.

Manufactures and Suppliers

Health and safety in manufacturing is also very important. For example: Corus manufacture steel which has many dangers. Their health and safety policy is lead by example as they believe every person working for them is responsible for their own health and safety and should set an example for others. Identify hazards and risks and ask your self is it worth the risk, Corus believe that understanding is the key to safe behaviour and they insure that their employees are trained and professionally skilled for their jobs. Corus also carry out incident analysis and prevention, all work related incident and near misses are reported and investigated to help stop a recurrence. A sharing and learning process is enforced, this means everyone in Corus is responsible for reporting hazards and learning from near misses, this will help prevent accidents. Corus also monitor, audit and review health and safety, they regularly conduct internal and external audits for their risk control measures and management systems. They also monitor behaviours at all levels to ensure they create a successful health and safety culture within their company.

A very good example – please reference your source in the text.

The CDM Regulations (2007)

The CDM regulations (construction, design and maintenance) aims to encourage everybody to work together and to make health and safety an integral part of the design, construction and management of projects. The regulations have been put in place to possibly identify and eliminate the hazards at the planning stage, to improve the overall health and safety throughout a project.

Jo, CDM is Construction Design and **Management**, not maintenance; although CDM does cover safe systems for maintaining buildings.

The CDM regulations state that the client must appoint a CDM co-ordinator. The client must also ensure that the construction phase of a project should not take place unless there are suitable welfare facilities are available. The client must provide the information in the health and safety file to the CDM co-ordinator and retain the file so it can be accessed. Also this file should be prepared and updated accordingly by the CDM co-ordinator.

You need to clearly stipulate who is the employer and employee here.

The CDM co-ordinator must advise and assist the client with his/her duties and notify the HSE (health and safety executive) of the project. The CDM co-ordinator must see through health and safety aspects of design work and co-operate with others involved with the project. Other roles for the CDM co-ordinator under the CDM regulations are to communicate between the client, design team and principal contractor.

The principal contractor must pass on sufficient information to subcontractors and prepare, develop and implement site rules before any work has started and subcontractors must obey and follow these rules. Also subcontractors should be made aware of any changes/updates made to the health and safety plan.

Management of Health and Safety at Work (1999)

The MHSAW (Management of Health and Safety at Work) are regulations to make more explicit what employers are required to do to manage health and safety and these regulations apply to everybody. The main requirements for employers are to carry out risk assessments. Employers with five or more employees need to record the findings of the risk assessment and make arrangements for implementing the health and safety measures identified as necessary. An emergency procedure is then put in place. Collective protective measure should be ensured so that the whole workplace is protected and those who work in it. To achieve this employers and employees sharing the same workplace must co-operate with each other regarding health and safety.

These specific regulations affect the CDM co-ordinator because he/she must ensure a health and safety file is prepared and all employees are made aware of it and follow throughout. The CDM co-ordinator must agree with the client and principal contractor to ensure health and safety training is provided to employees if necessary.

The principal contractor must ensure that the subcontractors are completing risk assessments before any work is being carried out. The subcontractors must be made aware of site rules and regulations such as: emergency procedures, assembly points and speed limits etc. This is the principal contractor responsibility to inform the subcontractors. The principal contractor must make the CDM coordinator aware if any subcontractors require health and safety training or information.

Subcontractors also must carry out risk assessments; they must notify the principal contractor of any unidentified risks. Also if an operative is not competent in carrying out a particular task, the principal contractor must be asked if any information or training is available.

You should have stated more information about the employee's duties here.

Keeping all the workforce informed of issues which can affect health and safety on site is a legal requirement under this regulation.

References

1. 'Introduction to Construction Health & Safety' by Ferrett & Hughes
2. www.hse.gov.uk
3. http://www.corusgroup.com/en/responsibility/health and safety/

ASSESSOR FEEDBACK FORM

STUDENT GRADING SHEET
Name: Jo Hussein
Unit 1 Assignment 1 Task 3 (P2)

You have researched these regulations well and have provided constructive answers. It is important to draw out the specific legal responsibilities of employers and employees for each regulation that you examined, but overall you have covered the issues to meet the criterion. I particularly like your example taken from the Corus information. To help you understand better, it would have been good to research similar examples for the other two regulations that you quoted.

Please quote the specific reference documents taken from the websites visited and include the date that you accessed them, not just the web address.

Jo, you have covered the issues well enough to meet grading criterion P2.

TASK 4 ANSWER

Unit 1 Assignment 1 Tasks 4a and 4b (P6)
Name: Jo Hussein
Identify and describe the role of the individual in accident recording and reporting procedures

Employers warned over dangers of forklift trucks after Berwick worker is paralysed

On 29th June 2006 Steven Rogers aged 29 of Berwick, sustained injuries after a downgrade bin which he was attempting to empty fell from the forks of a forklift truck and pinned him to the ground.

The incident occurred on a construction premises at Tweedmouth in Berwick upon Tweed and left Mr Rogers permanently paralysed. The penalty for this accident was that Silvery Construction Ltd was fined £20,000 after pleading guilty to breaching Health and Safety Laws. They were fined £16,000 for breaching Section 2(1) of Health and Safety at Work etc Act 1974 and £4,000 for Regulation 3(1) of the Management of Health and Safety at Work Regulations 1999. They were also ordered to pay costs of £5,397 at Berwick upon Tweed Magistrates' Court.

> This reads too much like the source itself. Aim to be more original in your work. A good technique to achieve this is to read the case study and highlight the key facts. You can then place these in a list and summarise the event in your own words. This means you rely more on the factual key points rather than the words of the author.

Following the court case the HSE reminded the construction industry that forklift trucks are a safety hazard and steps must be taken to safeguard against the dangers posed by these vehicles. HSE Inspector Martin Baillie said:

"Forklift trucks were responsible for just under 2,000 reportable incidents last year, including seven deaths. They are a potential danger to their operators and to other people in the vicinity if not operated with great care. Risks include being struck by a moving truck, crushed by an overturning vehicle, becoming trapped between a truck and an object or, as in this case being crushed by a falling load."

"Employers must ensure they assess the risks involved in any use of these vehicles and take appropriate steps to counter those risks. They must also provide adequate health and safety training for any employees operating forklift trucks."

This accident may have been prevented if the construction contractor ensured the load was adequately secured, carried out suitable risk assessments and ensured that their operators received adequate forklift truck training.

The role of the individual

Following an incident like this, all injuries, diseases and dangerous occurrences must be recorded and reported. This is as a result of various health and safety legislation, in particular the Reporting of Injuries, Diseases, and Dangerous Occurrences Regulations 1995 (RIDDOR). Such laws place a duty both on the individual and company. These will now be discussed in relation to the case study previously given.

> You state that we must record and report occurrences as a result of various laws but do not identify the key relevant legislation here (apart from RIDDOR). Always do this as it helps the assessor to see your awareness of the relevant issues.

General employees

Any person witnessing the event should have immediately reported the accident to a supervisor or manager. This would be to ensure that an employee with responsibility can take charge but also to confirm that an accident had taken place and for it to be addressed.

Anyone witnessing the incident must raise the alarm and inform someone with the authority to take control. Under no circumstance must someone try to interfere with the injured person or the accident scene itself. This is because it can cause further injuries to those involved and could also affect the site's layout which in turn could destroy evidence.

> How can an alarm be raised; do you mean in person or by telephone? You should make methods of informing people clearer here.

All witnesses are required to make a statement in order to help any investigation that may follow. Any information they provide must be to the best of their knowledge and should cover what they saw and what has happened. Depending on the situation they must also contact the emergency services to help deal with the incident and those injured.

In the event of this accident the following measures would have been expected. These would have been primarily controlled and managed by either a supervisor or manager:

1. All activities surrounding the incident would have stopped; workers should also have been moved away from the scene. This is to prevent the injury escalating and even more injuries occurring.

2. Medical attention for the injured would have been sought. This is to help the injured person as quick as possible.

3. Call the required emergency services. In this case study the fire services were needed to help release the casualty from the wreckage.

First Aiders

If an employee is competent and trained in first aid they can administer help. Some treatments must not be administered by a first aider and they should ensure they only apply what they've been trained to do.

The Injured Party

The injured person (depending on the severity of their injuries and whether they're able) will need to complete an accident report form. In the past this used to be part of an accident book but as these cannot be kept on site due to data protection this information must be sent directly to a manager who must send this to head office to be stored and recorded.

The accident report form includes the details of the person involved in the accident and also about the person filling in the accident form. Other information includes the date of the accident, its time, how it happened and the injuries sustained. When the form is complete the person filling out the form signs it.

Supervisor or Manager

Once they have been informed the supervisor or manager must take control. This will ultimately depend on their position of authority but they will have a responsibility to manage the accident that has taken place. This could involve contacting the affected party's family, calling emergency services, taking witness statements and organising other workers so not to further affect the incident site. Witness statements must be taken from the affected parties. It is important to remember that victims and witnesses could include visitors and members of the public.

Any witness statements that are given should be checked for accuracy and details. This is important as people can forget important information over time so any supervisor or manager must ensure that details provided state the witnesses' details and a detailed account of what they experienced.

The supervisor or manager must also check the victim's information for accuracy and its presence too. All data requested must be given and checked; for example day, time, injuries etc.

Discussing the importance of information accuracy is good. You could have referred to how this promotes better safety in the long term.

Periods of reporting vary and so do the methods. Reflection of this should be shown here. Refer back to the HSE website for more information.

When they have gathered this data they must report this to their company's Health and Safety department or individuals accountable for this. This is always dependant on the size of the company but in every case all information must be gathered. This is so it can be reported to the RIDDOR Incident Centre within the time frames RIDDOR states; normally 10 days of the accident happening.

Tutor feedback

Managers and employees in charge of safety must review the accident and ensure that all requirements of RIDDOR are met. This means that they will need to complete further documentation to evidence the findings of their subsequent investigation. This is in order to demonstrate that an investigation has taken place and that measures have been put in place to prevent further accidents. All this information must be added to the Health and Safety file for evidence.

Following this and other accidents on site the company's Health and Safety committee/board or senior management will meet to discuss the safety of operations on site. This is to review whether some events on site are due to poor planning and control measures. If this is the case the company may highlight that blind spots, inadequate risk assessments or poor methods exist on site. In other words, as a result of this review the company may be able to prevent similar accidents or events happening in the future. It is important to note that every accident will need to be investigated individually to assess the cause of the accident and to see if it can be prevented in future.

Though not evidenced by this case study it is also important that managers report any dangerous occurrences or near misses that occur on site. This is again to ensure that action can take place to prevent these occurrences and near misses from happening in the future.

Ultimately, it is important that all information is gathered correctly as any incident that is taken to court will need evidence to either enforce or prevent prosecution. Witness statements, accident logs and investigation reports etc are all legally required; however to aid this process, photographs can help to accurately document the scene.

Interesting overview of how data collection can help the company improve its health and safety. You could have referred back to the question and the role of the individual here. You could have also stressed how this information gets transferred between various levels in the company's hierarchy and why it's important to communicate any significant findings.

You should have evidenced what a near miss and dangerous occurrence is here.

ASSESSOR FEEDBACK FORM

STUDENT GRADING SHEET
Name: Jo Hussein
Unit 1 Assignment 1 Task 4 (P6)

You have presented the information clearly and logically and used appropriate language to make the document easy for the reader to follow.

You have broken down the information into clear sections and this demonstrates you are aware of those with responsibility following an accident on site.

Throughout your answer you refer to the relevant data and procedures, and the reasons for collecting or using this data. In places you should have given more information regarding its relationship to RIDDOR and improving company safety. However, what you consistently do is state what should be done and who should do take the appropriate action.

In this response you have not used images or references to support the comments you make. Images can help clarify the points you're making. References on the other hand are vital to demonstrate that sources you are using are accurate and up to date. In future you must do this.

As you have a good general understanding of the role of the individual in accident recording and reporting, **you have achieved grading criterion P6**.

TASK 5 ANSWER

Unit 1 Assignment 1 Tasks 5a, 5b and 5c (M3)
Name: Jo Hussein
Explain how collecting accurate data and information on accidents and incidents contributes to improvements in health, safety and welfare in the workplace

Part A

Every day on construction sites in the UK there are accidents occurring. These accidents can result in minor cuts and bruises to injuries which leave workers fighting for their lives and others that are ultimately fatal. Collecting information about these occurrences is vital as it helps the industry to analyse what is occurring and put in place measures to improve the safety of its workforce.

The process of recording and analysing data can sometimes be difficult as it requires workers to provide all detail requested. If people are not supervised properly or lazy in their work, information can be lost and this can affect the quality of control. Not only is collecting accurate data important for the company itself, it is also required by the HSE too so that they can manage safety effectively across the UK.

By collecting accident data in the form of an accident and incident report a company is able to gain a better picture of what is occurring on their construction sites or in their workplaces. Such data provides information about the nature of the incidents, why they happened and who was involved. For a company this may highlight whether a particular member of staff needs training or whether the whole workforce needs this too. It may also indicate whether poor control measures or management exist and whether new resources must be provided. Only by revealing and noticing this requirement can companies put in place measures to improve health, safety and welfare in the workplace.

Health and safety data is not only important for a company; it also helps those managing it throughout this and other industries. In other words, the collection and analysis of such information helps the HSE improve the safety in the construction industry. This is because through RIDDOR and other mechanisms (for example court cases) the UK and European industry is able to gather statistics. With such great quantities of information analysis can then investigate whether accidents relate to specific occupations, age groups, gender or location.

On this basis the HSE can establish whether new laws need to be created or initiatives need to be started in order to prevent further accidents occurring. For example, if the majority of accident data indicated that accidents resulted from falls from height, the HSE would then try to find a solution to address this problem. This could include changing laws regarding ladders as what happened in the past.

Part B

In recent years the number of falls from height has decreased. This is due to new laws and regulations set in place by the HSE reviewing accident data. In 1996/7 the fatal count of falls was around 90 workers per year and it approached 2004/5 they fell dramatically. They now stand at 50–55 workers per year. This is almost half in fewer than 10 years.

As stated above new changes in law meant that the number of fatalities fell. However it was the specific requirements within this law that made this possible, for example ensuring that safety equipment must be worn whilst working at height, toe boards and railings must be present, and ensuring that alternative procedures such as scissor lifts are used. All of these came as a result of being aware of what led to the number of fatalities and putting in place legal requirements to address this.

Good general introduction relating back to the task itself. Always use introductions in your work.

Problems relating to accurate data collection are discussed here. This is a relevant point to make. State what can be done to improve this process and how data findings can be communicated effectively.

Good general discussion showing how data collection can help various bodies within the industry. You should have referred to actual legislation by name and date.

Your answer is missing the degree of data required to meet this criterion. It could be strengthened in the following ways:
- Include more accident data.
- Use graphs and figures to state your findings.
- Regularly refer to any patterns and findings.
- Use your statistics to prove your point.
- Regularly refer back to the task and criterion.

**Tutor
feedback**

Part C

On construction sites the majority of workers tend to be men and as a result a larger number of workers injured are male. Women workers can be and are injured too on construction sites, however both male and female workers have the same responsibilities by law and thus the same role to play in protecting themselves and others.

Your point is not very clear here. Rewrite this to show how it relates to the criterion and task set.

It has already been mentioned how data analysis can help to create improvements in health and safety but sometime this data could be taken for granted. For example, more men are likely to injure themselves as a result of lifting objects on site. This is not because men are more susceptible due to their build, but instead because there are a larger percentage of male workers on site. If those analysing data did not take this into account then they would target male workers unnecessarily. Instead as the injuries through lifting can affect both sexes this issue affects everyone and the data must be used to ensure equality in measures used to address this.

Age can play a part in creating health and safety issues on site. This could be because a junior who is new to the construction industry would be more likely to be injured due to their unfamiliarity with hazards on a construction site and also the ways in which they should behave. An older worker on the other hand would be more aware of hazards on a construction site. However, they on the other hand may be more complacent and therefore put themselves at risk as they have a 'can do' attitude. Furthermore the industry does expect them to build up more knowledge of safety precautions as they may have witnessed many dangerous situations in their past. Such experience means less risk should exist and looking at data of age groups hurt should indicate what needs there are to improve safety for all age groups.

You are exploring the various reasons why data can vary and also why accidents occur as a result. What you need to do is relate this information back to the question. Concentrate on demonstrating how data collection of this type can bring about changes in health and safety. Include statistics and data from each of these areas and then explore how these could bring about future changes in legislation and workplace methods.

There are many types of injury which can result from accidents on construction sites; these include minor, major and fatal injuries. Minor injuries can be seen as cuts, sprains, bruises but must keep a worker off work for 3 days. Major injuries would be classed as broken or fractured bones, dislocation of the shoulder, hip, knee or spine etc. A fatal injury would be any wound which leads to death – this could be a heart attack, fall, severe knock on the head or being run down by heavy plant machinery.

The three categories of injury help to make improvements in health and safety in the industry as the breakdown shows what injuries are more likely and also the extent of these happening. Using this data the industry can decide whether more needs to be done to stop the more significant injuries occurring on site or whether they need to put in place measures to stop minor wounds occurring. Either way, this data helps to identify the problems that exist and what can be done to address them.

The cause of an accident is a main factor to review. If data indicates that the same cause appears regularly then rules and regulations will need to be changed, created and implemented. As some accidents can cause more harm and injury than others it is also wise to know the accidents which cause the greatest harm and injury so they can be avoided.

It is always essential to include a conclusion in your essay writing. Include a piece of writing looking back over the key points you've made in this answer. A good tip is to print out this information and highlight your key points. You can then use these to write a summary to prove what your research found.

171

Tutor feedback

ASSESSOR FEEDBACK FORM

STUDENT GRADING SHEET

Name: Jo Hussein

Unit 1 Assignment 1 Task 5 (M3)

This answer shows a good general understanding of how data collection can lead to improvements in health and safety. This is evidenced by your general discussion and also the way in which you have logically structured your answer.

At this stage you have not provided enough evidence to gain the criterion fully. This is because you are not using statistics and data to support the comments you make. You need to include this information and then use it to support your points. For example include data showing the types and numbers of injuries that have occurred over the last 5 – 10 years. Then state what this may prove and discuss what the industry may need to do to address this.

Include as many graphs and charts that you can to evidence the ideas you have. You also need to ensure you use up to date data. This can easily be found using the HSE website. Keep referring back to the question and the criteiron itself. Ensure you state how this data analysis and collection will bring about future changes.

Grading criterion M3 is not yet achieved.

TASK 6 ANSWER

Unit 1 Assignment 1 Task 6 (D2)
Name: Jo Hussein
Evaluate a provided accident report and suggest improvements that could be made to workplace systems in the future to avoid a recurrence

This construction site's current management and evening procedures must be revised and changed immediately; this is because it could have easily escalated into a more major or even fatal accident.

It is quite clear that if Mr Brownsmith was not found by the visiting security guard then he could have died due to loss of blood. This means that new measures and changes need to be put in place to improve the safety of this workplace.

There are many improvements that could be made to the existing workplace. Necessary ones are given below:

- Covering
- Lighting
- Fencing
- Paving
- Changes to existing safety measures and management
- Better communication methods

When the site is just about to be locked up there are many things that could be done to prevent accidents from happening. When the workers are just leaving they could cover the foundations or drainage with wood or metal sheets, this would then prevent any workers falling into the excavations. This is however only a temporary measure as they should have been clear walkways and restricted areas to prevent workers going near excavations.

It is important to note that Mr Brownsmith was not injured from the actual fall but the reinforcing bar. This could have been prevented if they had been covered using plastic caps or rubber covers.

When Mr Brownsmith was locking up his accident could have been prevented if there were flood lights overlooking the whole of the site. This is because Mr Brownsmith would have then seen where he was going and could have avoided the excavation. This lighting measure would not only help during the evening lockup but also throughout the day when weather is bad too.

When the workers are also packing up to leave the site they could put some kind of fencing around the perimeter of the excavated areas. This could include moveable fencing or even rope. This could then prevent anyone from falling into the excavations. This could be the most minimum precaution that could be used to prevent the event from happening again. There could also be a designated route or path created around this and other areas of the site. This would mean that those manoeuvring by foot are directed to safer areas of the site.

To prevent the incident from happening again the company should enforce improvements to the lock up method. This would come as a result of an effective risk assessment that realises that this dangerous site needs more than one person to carry out this role effectively. A method statement would therefore require those locking up to have a minimum of two people present whether it's day or night. This could prevent any accidents from happening because there would be another person to look out where they are going, meaning that there would have a less chance of falling into the excavations.

An interesting starting statement. An introduction relating to the task should have been used first though.

These bullet points could have been listed in order of importance. Showing what measures should be implemented first would help the assessor to recognise what measures you would make a priority.

It is important to consider how measures may and do vary depending on the time of day.

Images would help here and throughout this answer.

Weak understanding shown here as you state 'some kind of'. You should make clear suggestions and be specific about the appropriate methods. Please note that rope is not an effective control method. Your answer is improved by using professional terms such as 'manoeuvre' and 'designated'.

To prevent this incident from happening again better site management and control measures are needed. New systems could include new rules such as one enforcing a two people lock up process. A more effective one would be to use the method statement to list the order in which locking up activities take place. A example of this would require the site to be locked up in order, therefore:

1. Back gate (drop off area)
2. Storage area
3. Holding area (materials)
4. Side gate
5. Main gate

This then means that if anything happens and they need to know where managers are on site, then they can look at the system and locate where they could be. This would also indicate the most logical and safest route for security guards, emergency workers or other staff to take.

There are also many ways in which communications could have prevented Mr Brownsmith waiting for the security guard to find him. Suitable methods would have included a walkie talkie, a mobile phone or even a loudspeaker. A walkie talkie would have been suitable as it can be used over a long distance and those using them can be heard clearly. A mobile phone would have been suitable as it would again create direct communication. Problems associated with phones include no network coverage and the need to have your colleagues' phone numbers. A loudspeaker could also have helped due to its ability to raise an alarm quickly. However as they are bulky and unheard if used on a large or even noisy site it means they are the least suitable suggestion here.

In conclusion every site should be aware of the risks and hazards they contain. In order to make it a successful working environment, hazards like this need to be tackled. It is lucky that the person only received a minor injury from the situation, as if it was a fatal accident it could have resulted in investigations and court proceedings; unnecessary when the measures given here are easy to enforce and implement.

174

You've stated measures that can be put in place on a daily basis. State what procedures should be carried out to investigate and establish why this incident occurred.

You could have referred to the need to create a safe site to prevent vandals breaking in and causing damage to themselves and your site. Remember, if trespassers are injured on your site you will be prosecuted unless suitable measures have been taken to prevent them gaining access. What about phoning emergency services directly? Consider this.

Some evidence of evaluation shown here. Try to do this throughout by showing why your suggestions could make a noticeable impact.

Good clear relevant conclusion.

ASSESSOR FEEDBACK FORM

STUDENT GRADING SHEET

Name: Jo Hussein
Unit 1 Assignment 1 Task 6 (D2)

This is a good answer that details all the ways in which the site could improve to avoid a reoccurrence of such an incident in the future.

You have broken your answer down to show how different methods can have an effect. What you have not done is show how these compare with each other and evaluate which methods would be more effective and why.

At present you are close to meeting the criterion but need to show more evaluation. You need to consider which of the methods you are suggesting will have most effect. You can do this by referring to associated factors such as cost, time and management.

Your discussion of a safe system of work is very interesting and shows that you have researched the correct methods. Compare this to your suggestion of communication methods. What would you put as a priority? Compare other control measures too. What is better and why?

Do not forget that site safety is always very important, not just because someone has been injured. Signage should always be in place, tool box talks should be used regularly and a site should be safe enough to work in. This is because a HSE inspector can visit and close the site down or issue a prohibition notice until safer measures have been put in place.

Remember that all of this can ultimately cost time and money for any company involved.

To gain the full criterion you must fully weigh up the suggestions you're giving and show what control measures should be made as a priority (because they will have greater impact). Make more links, comparisons and conclusions to show what is better.

Grading criterion D2 is not yet achieved.